Formal System Verification

Rolf Drechsler
Editor

Formal System Verification

State-of-the-Art and Future Trends

Springer

Editor
Rolf Drechsler
DFKI
University of Bremen
Bremen
Germany

ISBN 978-3-319-57683-1 ISBN 978-3-319-57685-5 (eBook)
DOI 10.1007/978-3-319-57685-5

Library of Congress Control Number: 2017938317

Printed on acid-free paper

This Springer imprint is published by Springer Nature
The registered company is Springer International Publishing AG
The registered company address is: Gewerbestrasse 11, 6330 Cham, Switzerland

To Fatma Akin

Preface

For more than four decades the complexity of circuits and systems has grown according to Moore's Law resulting in chips of several billion components. While already the synthesis on the different levels from the initial specification down to the layout is a challenging task, for all the individual steps the correctness has to be considered.

In the past, classical approaches based on simulation or emulation have been used. But these techniques do not scale well and reach their limits. Correctness can only be ensured by the use of formal methods. These techniques were proposed more than 30 years ago in the context of circuit and system design, and in the meantime exist very powerful tools that are used in industry for specific tasks, like the equivalence check of netlists on the Register Transfer Level (RTL).

But with increasing complexity of the systems there is a high demand for tools that are better scalable and also consider modeling beyond plain digital circuits. In this context analog and mixed signal circuits have to be included on the lower level, but also hardware-dependent software towards the higher levels of abstraction.

In this book, these advanced topics of using formal verification along the design flow with a special focus on the system level are addressed. World's leading researchers have contributed chapters, where they describe the underlying problems, possible solutions, and directions for future work.

The chapters in the order as they appear in this book are:

- *Formal Techniques for Verification and Coverage Analysis of Analog Systems* by Andreas Fürtig and Lars Hedrich
- *Verification of Incomplete Designs* by Bernd Becker, Christoph Scholl and Ralf Wimmer
- *Probabilistic Model Checking: Advances and Applications* by Marta Kwiatkowska, Gethin Norman and David Parker
- *Software in a Hardware View: New Models for HW-dependent Software in SoC Verification* by Carlos Villarraga, Dominik Stoffel and Wolfgang Kunz
- *Formal Verification—The Industrial Perspective* by Raik Brinkmann and David Kelf

On the different abstraction layers it is shown in which way formal methods can assist today to ensure functional correctness. The contributed chapters cover not only the latest results in academia but also descriptions of industrial tools and perspectives.

Bremen, Germany
June 2017

Rolf Drechsler

Acknowledgements

All contributions in this edited volume have been anonymously reviewed. I would like to express my thanks for the valuable comments of the reviewers and their fast feedback. Here, I also like to thank all the authors who did a great job in submitting contributions of a very high quality. My special thanks go to Daniel Große and Jannis Stoppe from my group in Bremen in helping with the preparation of the book. Finally, I would like to thank Nicole Lowary and Charles Glaser from Springer. All this would not have been possible without their steady support.

Bremen, Germany
June 2017

Rolf Drechsler

Contents

Editors and Contributors

About the Editor

Rolf Drechsler received the Diploma and Dr. Phil. Nat. degrees in Computer Science from J. W. Goethe University Frankfurt am Main, Germany, in 1992 and 1995, respectively. He was with the Institute of Computer Science, Albert-Ludwigs University, Freiburg im Breisgau, Germany, from 1995 to 2000, and with the Corporate Technology Department, Siemens AG, Munich, Germany, from 2000 to 2001. Since October 2001, he has been with the University of Bremen, Bremen, Germany, where he is currently Full Professor and the Head of the Group for Computer Architecture, Institute of Computer Science. Since 2011 he is also the director of the Cyber-Physical Systems group at the German Research Center for Artificial Intelligence (DFKI) in Bremen. His research interests include the development and design of data structures and algorithms with a focus on circuit and system design.

Contributors

Bernd Becker Institute of Computer Science, Albert-Ludwigs-Universität Freiburg, Freiburg im Breisgau, Germany

Raik Brinkmann OneSpin Solutions, Munich, Germany

Andreas Fürtig University of Frankfurt, Frankfurt/Main, Germany

Lars Hedrich University of Frankfurt, Frankfurt/Main, Germany

Dave Kelf OneSpin Solutions, Munich, Germany

Wolfgang Kunz Department of Electrical and Computer Engineering, University of Kaiserslautern, Kaiserslautern, Germany

Marta Kwiatkowska Department of Computer Science, University of Oxford, Oxford, UK

Gethin Norman School of Computing Science, University of Glasgow, Glasgow, UK

David Parker School of Computer Science, University of Birmingham, Birmingham, UK

Christoph Scholl Institute of Computer Science, Albert-Ludwigs-Universität Freiburg, Freiburg im Breisgau, Germany

Dominik Stoffel Department of Electrical and Computer Engineering, University of Kaiserslautern, Kaiserslautern, Germany

Carlos Villarraga Department of Electrical and Computer Engineering, University of Kaiserslautern, Kaiserslautern, Germany

Ralf Wimmer Institute of Computer Science, Albert-Ludwigs-Universität Freiburg, Freiburg im Breisgau, Germany

Chapter 1
Formal Techniques for Verification and Coverage Analysis of Analog Systems

Andreas Fürtig and Lars Hedrich

1.1 Introduction

Besides the pure digital and software-related verification methodologies, analog and mixed signal circuits and systems are also in strong focus of adding formal verification into the verification flow. In fact, the validation demand can be higher than in pure digital design because the system's behavior due to the continuous nature of the signals gives more freedom to signal processing. The main driving forces in IC and embedded systems market are communication and automotive circuits. Both demand for substantial analog parts coupled with digital parts. On top of that, these mixed-signal systems are connected with sensors and actors to the physical world in a control loop such that the overall system validation has to take the digital part, the analog part, the sensors, and the physical world into account. The analog parts (around 10–30% of the chip area) have fewer transistors (100–1000 for a block) and may add up in tens to hundreds of blocks. Clearly, these analog blocks are nonlinear dynamic circuits with the nonlinearity being a major property of the circuit.

The variety of analog circuits directly leads to a large variety of verification methods—even in simulation-based approaches. There are a lot of simulators on many levels of abstraction, starting from device simulators up to system-level design-languages and simulators like SystemC-AMS [1] and some specialized simulation algorithms e.g. for RF simulation [2], automatic behavioral model generation [3], or reliability modeling and simulation [4].

Unfortunately the analog/continuous circuits do not have a nice Boolean abstraction layer, which paves the way for digital formal verification tools. Hence, the available analog verification tools are more or less focused on the abstraction layer they reside on. Most tools have big complexity issues keeping the size of verifiable

A. Fürtig (✉) · L. Hedrich
University of Frankfurt, Frankfurt/Main, Germany
e-mail: fuertig@em.cs.uni-frankfurt.de
L. Hedrich
e-mail: hedrich@em.cs.uni-frankfurt.de

R. Drechsler (ed.), *Formal System Verification*, DOI 10.1007/978-3-319-57685-5_1

circuits small and hence adding more pressure to have dedicated tools for many abstraction layers. Furthermore, this demands for cross abstraction-layer tools like equivalence checking. We will present some ideas to formal verification on transistor level and extend them to close the gap to higher level abstractions.

Another way of increasing the confidence into the designed analog circuitry is the introduction of coverage measures. Depending on the type of coverage, one can get a closed form, well defined, formal coverage definition accompanied by proper algorithms. On the other end, there may be ad hoc methods with created tests from experienced designers having all to be fullfilled to get 100% coverage. In this chapter, we will describe and discuss some methods for analog coverage.

Coming to system level the picture turns into something slightly different. Here, a long tradition of hybrid system formal verification methods exists, which now has to be connected with the lower level circuitry. We will present some methods and their applications and show a way to close the gap down to transistor level using a chain of formal verification steps and a stack of abstracted behavioral models. We will show, how a system-level formal verification could be applied on an example consisting of a cyber-physical system.

This chapter begins with a state-of-the-art section, a description of the analog/-continuous state space and its algorithmic handling. We will then discuss several verification methods in Sect. 1.4 based on the analog state space, as well as an introduction to an analog coverage methodology (see Sect. 1.5 ff.). Section 1.8 combines all presented methods to apply formal verification to the system level.

1.2 State of the Art

As described in the other chapters of this book, formal verification of digital circuits has a long tradition and is deeply integrated in design companies' verification flow. However, for analog circuits or continuous systems on system level a need for accurate and fast verification methods becomes more important as cyber-physical systems and Internet of things produces a lot of sensors, actors, and hence continuous-centric designs.

In any case, the objective of formal verification is to mathematically prove properties of a system, usually at design time. It can be distinguished between reachability analysis, model checking, and more specific equivalence checking. Model checking is mainly used for formally verifying specified properties that in particular relate to safety and liveness of a system. Equivalence checking, being able to prove the equivalence of two implementations could be used to build a chain of proof from lower level implementations up to abstract behavioral models. Reachability analysis is the little friend of model checking, as it allows an easy straightforward way of a safety proof. For hybrid systems on system level, this technique is used the most.

System Level

A first approach to move from discrete systems to continuous systems was based on hybrid automata and eventually evolved in the tool HyTech [5]. The approach mainly uses a set-based approach on linear or piecewise nonlinear models [6–8] in order to compute reachable sets which can be checked for hitting a forbidden region.

Later, the tools for hybrid systems evolve introducing polyhedra as data structures [9] and trying the first time to compute reachable sets of nonlinear electrical circuits. These are circuits with low complexity enabling the manual piecewise modeling of nonlinearities like a tunnel diode in an oscillation circuit. However, the piecewise technique can help to model switching analog circuits on an abstract level for verification. Widely used examples are Sigma-Delta AD converter [10, 11].

The used underlying data structures and computation models began to broaden from Petri nets [12] to interval [13], affine methods [14], zonotopes [15] to support functions [16].

Transistor Level

However, to be more accurate on lower levels of abstraction one has to incorporate the transistor which is really hard to model by piecewise-linear hybrid systems—a today's transistor model consists of hundreds of nonlinear equations. Additionally, the Kirchhoff's laws do not allow solving the nonlinear differential algebraic system (DAE) for getting an explicit ordinary differential equation (ODE) description needed by most hybrid-systems approaches. Consequently, some different methods evolve. One approach extends affine arithmetics with a Newton iteration to solve nonlinear equation systems [14]. However, this approach still needs some simplified transistor models [17].

Other approaches try to abstract the exact nonlinear behavior by determining an analog state transition graph using the original netlist and detailed BSIM transistor models with simulator engines and full SPICE-accuracy [18–21].

Formal Languages

In [11], the authors propose an assertion-based verification flow based on PSL. Other approaches extend CTL to be able to describe analog behavior and to check time constraints [22, 23]. The latter can be driven down to transistor level in order to ease usage by allowing the direct formulation of often used specifications, e.g., gain, PSRR, slew rate, and input/output voltage ranges [24]. All these methods can be used to model check the design under verification or at least enable an assertion-based verification (ABV) by running in parallel with a simulation. This is a big advantage, as it scales better with netlist size by not exploring the exponential growing state space. The disadvantage is that the proving character will be lost. All these formal language based approaches suffer hard to interpret results and the perennial "translate specification into a formal language" problem.

Coverage metrics

One traditional—comparable to assertion-based verification—direction to systematically check analog circuits may be the full automatic characterization [25] based on formalized specifications using machine readable specifications [26] or formal languages such as PSL [11]. However, the effort to setup these specifications is sometimes large and, even worse, they do not guarantee to find unknown bugs because they rely on predefined input stimuli for each performance test case. With simulation only, there still exist uncovered scenarios which may later pop up as a bug in the field.

Direct formal verification tools like mentioned above will certainly help as it can guarantee to find the problematic design flaws violating the specification. However, they often suffer from long runtimes, hard formulate specifications and hard to interpret results.

A compromise could be the use of coverage metrics and coverage increasing measures. The digital world has developed a lot of coverage metrics [27, 28] and uses them successfully. Depending on the complexity of the device under verification (DUV), the methods are more or less complete. The complete methods investigate, for example, the finite-state machine (FSM) [29] and have some means to try to restrict the simulation input stimuli to the relevant part of the state space (see SFSM in [29]). The less complete methods (code coverage, specification coverage) use measures to guide the verification to the most probable bug location for example by systematically visiting each conditional branch in an HDL-description.

For analog circuits, a very low number of coverage investigating approaches besides the above explained formal verification techniques exist. There are some approaches stemming from the test community measuring and increasing the analog fault coverage [30, 31]. However, they are not intended to find functional faults. [32] tries also to increase the confidence in the functional verification using a high-level functional model but without a systematic method to increase some underlying measure. Two other approaches are built for hybrid systems [33, 34], suffering from being able to handle strongly nonlinear analog circuits on transistor level.

In this chapter, we will give an overview to actual formal verification techniques for analog circuits and systems which come up with some of the mentioned problems. We will present a methodology to enable a cross-layer verification to close the formal verification gap from transistor level up to system level.

1.3 State-Space Description

All techniques to verify analog circuits and systems formally have to deal with the continuous state space spanned by energy-storing elements like capacitors and inductors, or other physical states like position, velocity etc. Especially when reaching system level these physical variables and states may be important. In general, we can incorporate them and start with an implicit nonlinear first-order differential algebraic equation (DAE) system

$$\mathbf{f}(\dot{\mathbf{x}}(t), \mathbf{x}(t), \mathbf{u}(t)) = \mathbf{0} \tag{1.1}$$

with an input vector $\mathbf{u}(t)$ and the vector of the system variables $\mathbf{x}(t)$ and its time derivative $\dot{\mathbf{x}}(t)$. We always denote vectors in upright bold-face. For verification tasks, an output variable $y(t)$ has to be identified. In general, this is done by defining an output equation.

$$g(\mathbf{x}(t), y(t)) = 0 \tag{1.2}$$

In most cases, the output variable is a system variable $x_i(t)$. In this case, g is a simple selection function in terms of $\mathbf{x}(t)$ and $y(t)$.

This equation system can be set up automatically by a modified nodal analysis (MNA) on transistor level or manually for higher abstraction levels. As explained in Sect. 1.2 many of the high-level tools use a system of ordinary differential equations (ODEs).

$$\dot{\mathbf{x}} = \tilde{\mathbf{f}}(\mathbf{x}, \mathbf{u}) \tag{1.3}$$

Equation (1.1) could in general not be converted into this explicit ODE form (1.3). In both cases, a state space with n_d dimensions is spanned by the system variables of the energy-storing elements $\mathbf{x}_e$ extended by the n_i input variables. For the ODE case n_d is equal to the number of variables n_x, leading to an extended state space dimension of $n_s = n_d + n_i$. In the DAE case, n_d is less or equal than the number of variables n_x due to algebraic equations and/or linear-dependent state variables. In most cases the number of algebraic equations is much larger than the number of independent state variables n_d, e.g., transistor-level circuits have around 40 times more algebraic equations than differential equations, often hidden in the transistor models.

For analog circuits with many parasitic energy-storing elements, the number of dimensions n_d can be further reduced. Many order reduction methods exist for linear or linearized systems, e.g., dominant pole [35] or Pade approximation [36]. In the nonlinear case, advanced methods have to be used to extend a linearized order reduction to the nonlinear state space [37]. However, both methods end up with a reduced, much smaller state-space dimension n_{d^*} which could be well estimated by the number of wanted poles, often known by the analog designer. For the ease of reading we will always use n_d, even if we use an order reduction method resulting in n_{d^*} states.

1.3.1 Solving a DAE System

Without loss of generality, we can assume that we can handle the following also for ODE systems. The general working horse of an analog circuit designer is a transient simulation. It has the highest accuracy. Mathematically, this transient simulation is the solution of an initial value problem (IVP) defined by a given starting state $\mathbf{x}_S$, a given input stimuli $\hat{\mathbf{u}}(t)$ and the DAE systems:

$$\begin{aligned} \mathbf{f}(\dot{\mathbf{x}}(t), \mathbf{x}(t), \hat{\mathbf{u}}(t)) &= \mathbf{0} \\ g(\mathbf{x}(t), y(t)) &= 0 \\ \mathbf{x}(0) &= \mathbf{x}_S \end{aligned} \tag{1.4}$$

The solution $\hat{\mathbf{x}}(t)$, $\hat{y}(t)$ defines a trajectory $\mathfrak{T}_{\mathbf{x}_S,\hat{\mathbf{u}}(t)} = \{\hat{\mathbf{x}}(t), \hat{y}(t)\}$ through the state space starting at $\mathbf{x}_S$ (see Fig. 1.2). DC and AC analysis are special simulation cases, which can be covered by the transient analysis. Also harmonic balance, shooting methods, or periodic steady-state methods are special simulation cases, mainly to save simulation time. In principle, all validation results could be obtained by a transient simulation. A validation scheme today requires a set of simulations (hence a set of input stimuli $\hat{\mathbf{u}}_i(t)$) and calculates a set of trajectories $\mathfrak{T}_{\hat{\mathbf{u}}_i(t)}$. If a specification fails, one of these trajectories will fail and will result in a counterexample.

1.3.2 Analog Transition System

Subsequently, the DAE system shall be transformed into a discrete state space model M_{ATS} as defined in Definition 1.1 by application of a trajectory-directed discretization algorithm [38]:

$$\mathbf{f}(\dot{\mathbf{x}}, \mathbf{x}, \mathbf{u}) = \mathbf{0} \xrightarrow{\text{discrete modeling}} M_{ATS} \tag{1.5}$$

Definition 1.1 Analog Transition System (ATS)
For the ATS we define a five-tuple $M_{ATS} = (\Sigma, R, L_V, T, L_\lambda)$ where

- Σ is a finite set of states of the system.
- $R \subseteq \Sigma \times \Sigma$ is a total transition relation, hence for every state $\sigma \in \Sigma$ there exists a state σ' such that $r = (\sigma, \sigma') \in R$.
- $L_V : \Sigma \to \mathbb{R}^{n_d}$ is a labeling function that labels each state with the vector of n_d variables containing the values of the state space variables and the inputs of the DAE system.
- $L_X : \Sigma \to \mathbb{R}^{n_x}$ is a labeling function that labels each state with the vector of n_x variables containing the values in this state of the inner variables $\mathbf{x}$ of the DAE system.
- $T : R \to \mathbb{R}_0^+$ is a labeling function that labels each transition r from σ to σ' with a real-valued positive or zero transition time that represents the time required for the trajectory in the state space between these states.
- $L_\lambda : \Sigma \to \mathbb{R}^{n_\lambda}$ is a labeling function that labels each state with a vector of the n_λ eigenvalues associated with the state.

Within the structure M_{ATS}, a path π beginning at state σ is a sequence of states $\pi = \sigma_0, \sigma_1, \sigma_2, \ldots, \sigma_n$ with $\sigma_0 = \sigma$ and $r_{i,i+1} = (\sigma_i, \sigma_{i+1}) \in R$ for $0 \le i < n$. An example of an ATS is shown in Fig. 1.1.

The discretization into an ATS is performed by using an extended simulator to first compute a consistent starting state vector $\mathbf{x}_s$, and second solve the DAE system

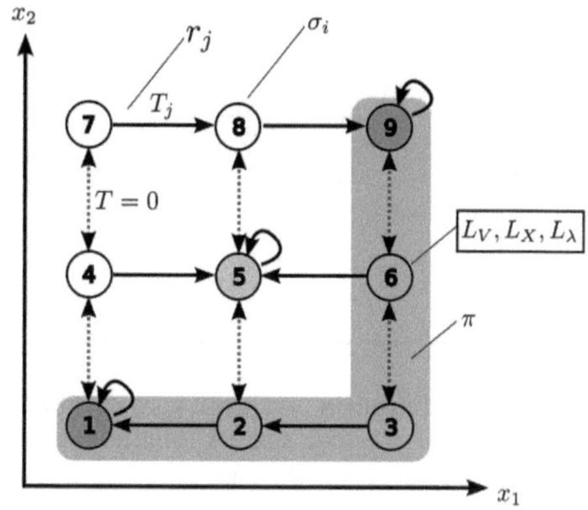

Fig. 1.1 Principle of an ATS: The system consists of states (denoted by numbers) and the transitions between those states. A transition indicates changes in the input to the system (*dotted arrows*) and timed transitions. The *gray* area marks *one possible* path π from state 9 to state 1 through the ATS

for a given time step Δt and a constant input vector $\mathbf{u} = const$ (see Fig. 1.2left). The resulting trajectory $\mathfrak{T}_{\mathbf{x}_S,\hat{\mathbf{u}}(t)=const}$ will lead to a new state σ_j and define the relation $r_{i,j} = (\sigma_i, \sigma_j) \in R$ and a discrete flow vector:

$$\Delta\mathbf{x}_{i,j} = \mathbf{x}_j - \mathbf{x}_i \tag{1.6}$$

Additionally, depending on the dimension of the reduced state space—remember, an order reduction technique is used—an orthogonal base set is generated using the Gram–Schmidt orthogonalization method. By addition/subtraction of the orthogonal set and the initial starting point, new starting points for transient steps are calculated for which in turn the orthogonal sets are constructed if not already existing.

For input dimension, this orthogonal base is also used. In that case, an input step can be performed in reality leading to a legal change in the reduced extended state space which is considered by introducing a relation with $T = 0$ time and oriented in both directions (see Fig. 1.1).

The time step Δt and length of the orthogonal steps is calculated comparing the length and angle of the system variables' derivatives $\dot{\mathbf{x}}$ in that region. The result should be a discretization such that the system's state space is divided parallel and orthogonal to the trajectories of the system dynamics into geometric objects representing areas of nearly homogeneous behavior.

The flow vectors $\Delta\mathbf{x}_{i,j}$ also define the transition relation of the geometric objects and the transition time. The DC operating points of the circuit are modeled with self-transitions to indicate that the system can stay here for any time period when the inputs are held constant. Accordingly, these self- transitions have no timing information (see Fig. 1.1, vertices 1, 5, and 9). For a detailed discussion on the soundness and accuracy of the discretization method, please refer to [38]. In order to calculate the eigenvalues of each state during the discretization process, the system's dynamics are linearized in the specific state and then transformed into the frequency domain using Laplace transformation. The number of nonzero entries in Kronecker's canonical form of the

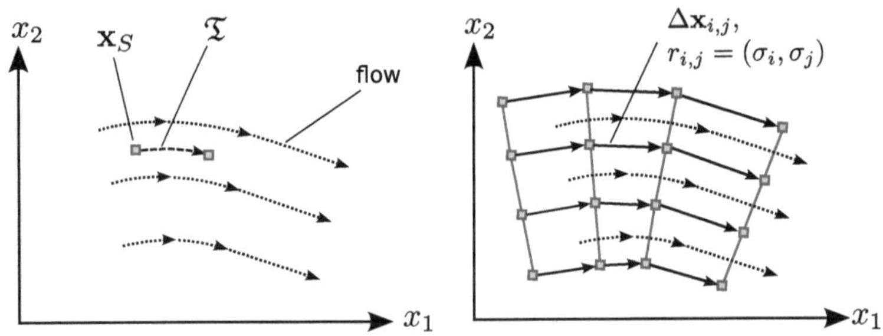

Fig. 1.2 Discretization of a phase portrait (flow): *Left* The flow and a discretization step starting from a point in the state space $\mathbf{x}_s$ and following a trajectory $\mathfrak{T}_{\mathbf{x}_S,\hat{\mathbf{u}}(t)=const}$. *Right* The resulting discretization based on orthogonal sampling and following the trajectories. Each connection element (*arrow*) represents both, the discrete flow vector $\Delta\mathbf{x}_{i,j}$ and the relation $r_{i,j}$

transformed capacitance matrix of the frequency domain representation corresponds to the number n_λ of eigenvalues in the generalized eigenvalue problem. For a detailed description of the eigenvalue decomposition, please refer to [37].

Finally, the transition system is mapped to the ATS, considering each hyper cell as a vertex of the graph with directed edges defined by the transition relation r_j with the corresponding transition times T_j. Each vertex of the graph is labeled with the state space variable values at the center of the corresponding geometric object L_V, the full solution vector of $\mathbf{x}$ at the point L_X and the eigenvalues L_λ.

While the complexity of the state-space modeling process is exponential in the number of energy-storing elements and inputs of a circuit, relevant analog circuit blocks usually do not exceed a system order of 8, which can be handled well by this approach. Moreover, by application of an eigenvalue-based model order reduction

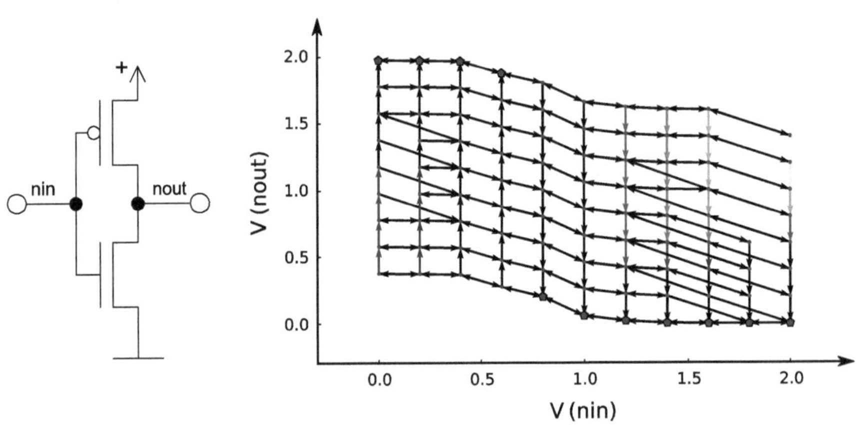

Fig. 1.3 Full discretization of an inverter circuit (*left*) and the resulting two-dimensional full M_{ATS} system. The DC operating points are shown in *blue*

of the DAE system [39], circuits with more than 200 parasitic capacitances can be handled. This is achieved by reducing the state space to the dominant state variables of a system and discarding the parasitic ones which will not affect the system behavior above a defined threshold in a given frequency range (Fig. 1.3).

The described modeling process is implemented as an extension to the industrial in-house simulator as well as using the public domain analog simulator GNUCAP [40].

1.4 Verification Methodology

A vast number of verification methods exist for analog circuits and systems depending on the type of circuits and their specifications. However, most of them do not include formal methods. In contrast, Fig. 1.4 introduces a flow of a verification methodology using different formal tools for the verification task in analog design. Starting from a circuit and system description on different abstraction levels, a discretization method calculates an ATS, or a direct equivalence-checking method calculates the difference between two implementations [37]. On the ATS, a model-checking algorithm can prove the properties of the given ASL, which is discussed later in this section. A coverage analysis and optimization method can calculate a coverage for given validation stimuli or optimize them to increase the coverage. With stimuli that assure a high coverage another equivalence checking by bi-simulation could also be possible.

In the following, we will discuss some methods in more detail.

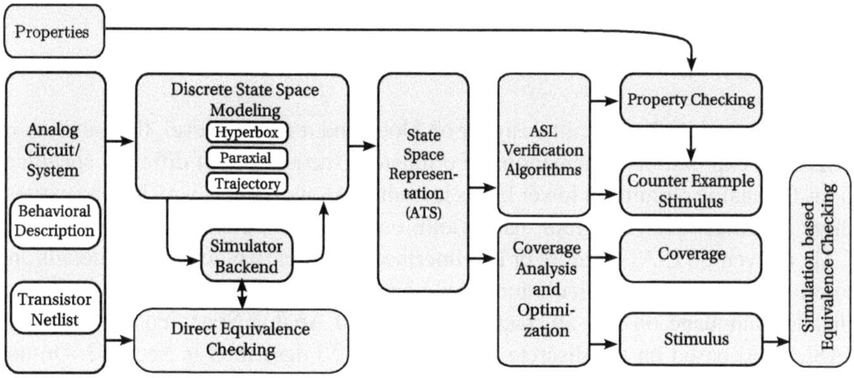

Fig. 1.4 Overview of a possible formal verification methodology

1.4.1 Model Checking

Once the ATS graph has been calculated, model-checking techniques could be applied to the resulting discrete model of our analog system. In fact, the discrete graph has some other semantics compared to the sequential digital circuit's case. As seen in Definition 1.1 (and also in Fig. 1.1), timing information T are available on each edge and information about the states' and other variables' values are available in each node. Both are important to verify analog properties as we will see later.

The first step to reason about continuous behavior is to introduce a comparison with a real number. In the ATS case, all variables stored as real numbers in the labels L_V, L_X, L_λ are possible candidates for that comparison $x_i < const$. Using that comparison, several sets of states could be defined as proposed in various papers [11, 22, 23, 41], enabling specifications over continuous values. To enable specifications over a continuous time, e.g., slew rate, settling time etc., a timed version of CTL or PSL should be used. A big step forward was TCTL [42], which was extended or combined with continuous values properties in the papers mentioned above. With all these languages available, model checking for more or less simple CTL/PSL like performance descriptions could be carried out. For example, a simple safety property avoiding bad behavior could be expressed as follows:

$$M \models AG\left(\neg\left(V_{out} > 5\right)\right) \tag{1.7}$$

This formula demands that an output voltage V_{out} does never exceed 5 V. In this simple case, a reachability analysis could also proof that formula.

1.4.2 Analog Specification Language (ASL)

For the general analog circuit designer on block and transistor level, the handling of PSL/CTL expressions is demanding. Fortunately, the number of different specifications for analog circuits on lower levels is limited. However, it is very hard to express these specifications (like gain, bandwidth, output swing, etc.) directly using PSL, CTL, or even CTL-AT. A more or less intermediate layer to hide the CTL details and to provide some macro calculations could help here. We proposed a property specification language on low abstraction level called Analog Specification Language (ASL) [43] based on the discrete state space (ATS) described in Sect. 1.3. On one hand, it enables the easy description of specifications like offset, gain, CMRR, PSRR, slew rate, overshoot, startup time (for oscillators, charge pumps), general oscillation properties like frequency, attractors, steady states, VCO input sensitivities, and some more. On the other hand, this language comes with extensions to CTL/PSL which allows a direct evaluation of analog properties like gain, time derivatives or time constants and stability properties for oscillations.

1.4.3 *ASL-Example: Verification of Oscillation and Oscillator Voltage Sensitivity*

In the following, we will exemplarily show how to specify complex analog properties with the Analog Specification Language (ASL), using an example taken from [43]. For voltage-controlled oscillator (VCO) circuits, we want to verify properties of the oscillation and the voltage sensitivity. In the time domain, the oscillation is represented by a periodic behavior of a circuit variable as shown in Fig. 1.5a. Transferred to the continuous state space, a cyclic path between at least two state-space variables can be identified as illustrated in Fig. 1.5b. This results in a set of states connected to a cycle in the discretized graph structure as shown in Fig. 1.5c. The simple check whether there is an oscillation within a defined oscillation period range between `%spec_min` and `%spec_max` in the considered circuit model can be formulated in ASL as follows:

```
assign(%oscillation_period_min, min) oscillation;
assign(%oscillation_period_max, max) oscillation;

for %oscillation_period_min assert [>= %spec_min];
for %oscillation_period_max assert [<= %spec_max];
```

For verifying the sensitivity relation between the control voltage and the oscillation frequency, we constrain the input voltage to different values. At each of these input voltages the oscillation frequency is determined. Comparing each two oscillation frequencies of consecutive input values to their input voltage difference, the sensitivity

$$K_{VCO} = \frac{\partial(\text{Oscillation Frequency})}{\partial V_{in}} \tag{1.8}$$

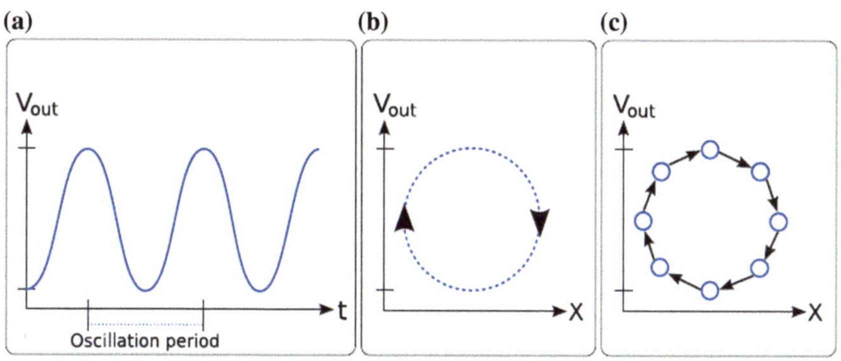

Fig. 1.5 Oscillation in the time domain (**a**), in the continuous state space (**b**), and in the discrete graph structure (**c**)

can be determined. In the following methodology, only two different input values are considered for the purpose of clarity. Considering more than two input values, the deviation between the calculated local factors K_{VCO} gives information about the linearity of the VCO. At first, the constrained input voltage areas in state space have to be assigned to set variables as follows:

```
inp_set_1 =
  value(V_in)[%inp_voltage_1-0.01, %inp_voltage_1
      +0.01];

inp_voltage_2 = %inp_voltage_1 + %inp_step;

inp_set_2 =
  value(V_in)[%inp_voltage_2-0.01, %inp_voltage_2
      +0.01];
```

On the selected state-space slices, oscillation periods are determined. Although the average oscillation period for a given input voltage is considered, this approach is also valid for the minimum or maximum oscillation period:

```
osci_set_1 = on inp_set_1
  assign(%osci_period_1, average) oscillation;

osci_set_2 = on inp_set_2
  assign(%osci_period_2, average) oscillation;
```

Subsequently, the amount of the oscillation frequency change and the input voltage change are determined:

```
frequency_delta =
  (1/%osci_period_2) - (1/%osci_period_1);

input_delta =
  %inp_voltage_2 - %inp_voltage_1;
```

In the final step, we assert for the sensitivity property calculated according to Eq. 1.8, that the relative error is within a percental range specified by the number variable `%tolerance` around the specified value `%K_VCO`:

```
for %frequency_delta / %input_delta
  assert [%K_VCO-%tolerance/2, %K_VCO+%tolerance/2];
```

As shown with this example, one is able to formulate complex ASL statements in time domain to check some frequency domain or mixed domain properties. For more details of ASL we refer to [37].

1.4.4 Model Checking of an SRAM Cell

As an example for model checking, we want to prove properties of a static RAM cell (SRAM). It has a nonlinear transfer and storage characteristic, where it is important to be non sensitive with respect to external disturbances, like adjacent capacitively coupled wires or particle strikes bringing a charge into the storing SRAM cell. A schematic of an SRAM cell is shown in Fig. 1.6. The schematic contains also an aggressor current source modeling the influence of an aggressor line or a particle strike.

With the discretization approach of Sect. 1.3, two state variables (node voltages of nodes "OUTP" and "OUTN") and one input variable (the aggressor current) span a 3-dimensional extended state space. In Fig. 1.7, two sub planes for $I_{Aggr} = 0$ and $I_{Aggr} = 0.0025$ are shown. The stable fix points for storing a "1" or a "0" can be clearly identified in the upper left and lower right corner of the left figure. Inspecting the right figure, for a large aggressor current I_{Aggr}, only one fix point is left over. Hence, a disturbance following a change of the stored value in the SRAM cell could be assumed.

To formalize the process of checking the correct behavior, the following expressions could be used, which could also be implemented in ASL:

$$
\begin{aligned}
high &= V(OUTP) > 1.7 \wedge V(OUTN) < 1.3 \wedge I_{Aggr} = 0 \\
low &= V(OUTP) < 1.3 \wedge V(OUTN) > 1.7 \wedge I_{Aggr} = 0 \\
\Phi_{slice} &= I_{aggr} > -0.0013 \wedge I_{aggr} < 0.0013 \qquad (1.9)\\
\Phi_{appr} &= \Phi_{slice} \wedge EF(high) \\
\Phi_{fail} &= \Phi_{appr} \wedge low; \\
M &\models AG\left(\neg\Phi_{fail})\right)
\end{aligned}
$$

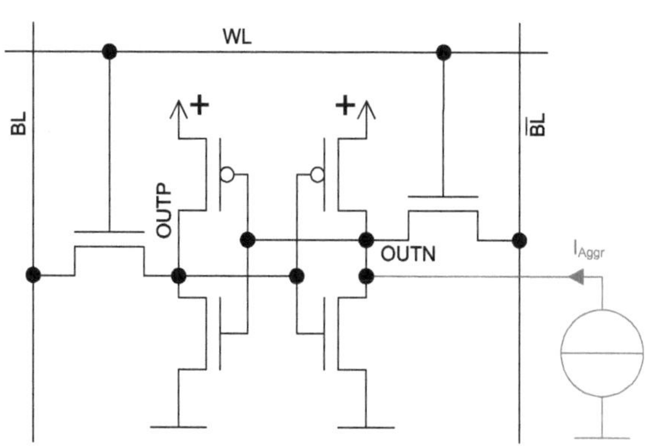

Fig. 1.6 Schematic of an SRAM cell with an aggressor current source

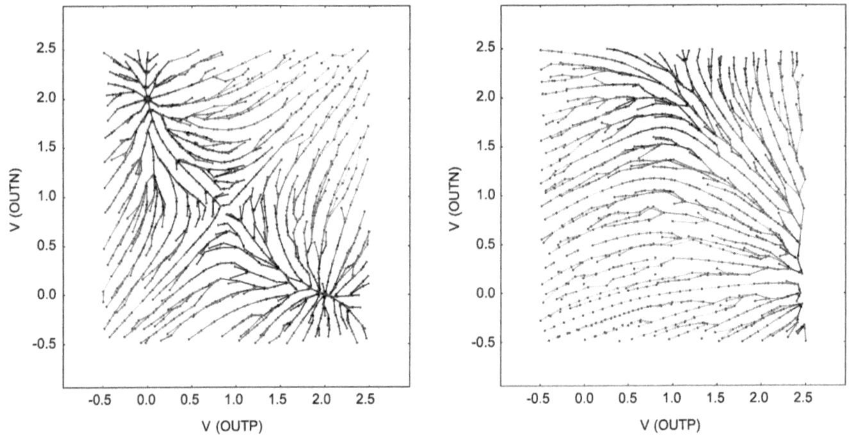

Fig. 1.7 SRAM: Discretized state space. The shown dimensions are the node voltages of nodes "OUTP" and "OUTN". *Left* cut out plane at $I_{Aggr} = 0$. *Right* cut out plane at $I_{Aggr} = 0.0025$

Fig. 1.8 Counterexample in the 3-dimensional state space for a bit flip with an appropriate aggressor current. The trajectory in the state space runs from $V(OUTN) = 0V$, $V(OUTP) = 2V$ to $V(OUTN) = 2V$, $V(OUTP) = 0V$, indicating the bit flip

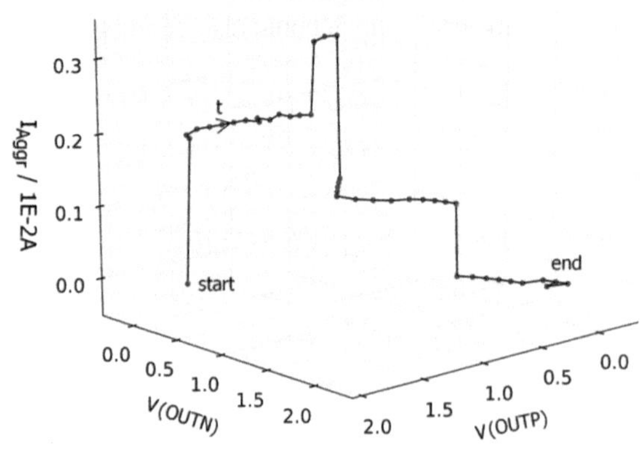

The model-checking process of these equations succeeds, indicating that aggressor currents between $-0.0013 < I_{aggr} < 0.0013$ are not able to flip the bit in the given SRAM cell. However, if we investigate in larger aggressor currents by increasing the Φ_{slice} to $-0.005 < I_{aggr} < 0.005$ then the above formula (Eq. 1.9) fails. A counterexample could be generated leading to the trajectory shown in Fig. 1.8.

Simulating the counterexample with a standard analog simulator confirms the failure resulting from the increased aggressor current I_{Aggr} (see Fig. 1.9).

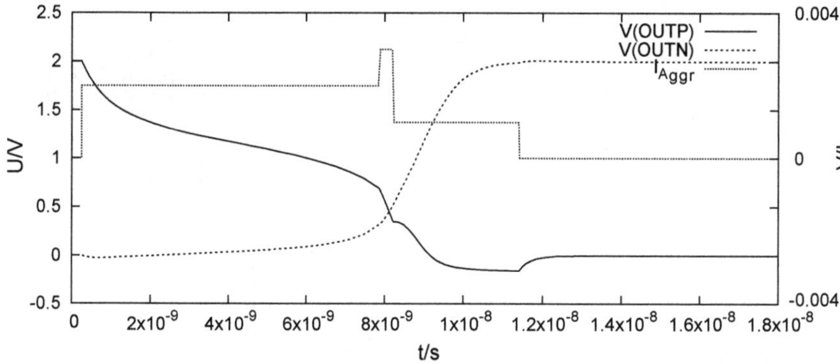

Fig. 1.9 Simulation result of the generated counterexample (the I_{Aggr} stimulus), clearly uncovering the bit flip

1.5 State Space Coverage

In this section, we introduce a state-space coverage metric and an algorithmic concept to maximize the metric by generating input stimuli based on path planning information obtained from an ATS.

We define a state space coverage ζ as the ratio between visited states during a simulation run and the sum of all reachable states Σ_R of a given circuit. In our case, the reachable states Σ_R are computed from all states Σ visited by the state-space discretization described in the previous chapter using a simple set based reachability algorithm. Our goal is to find a coverage measure, being able to compute for any given transient simulation response of a simulator a measure with the following properties:

- A high coverage value implies a high probability that all possible faults of the circuits could be detected.
- A high coverage value shows the designer that the created circuit was tested with a sufficient amount of test data.
- The measure has to be monotonic in the number of visited states: If more states are visited, the measure should increase.

1.5.1 State-Space Coverage Calculation

To calculate a coverage for a transient simulation result, we store the previously defined Analog Transition System M_{ATS} in a suitable space-partitioning data structure in form of a k-d tree [44]. The number of nodes in this tree equals the number of states in the system.

In a first straightforward approach, one can compute for each point s_i of a simulation response S the nearest neighbor using the k-d tree data structure. Then, C

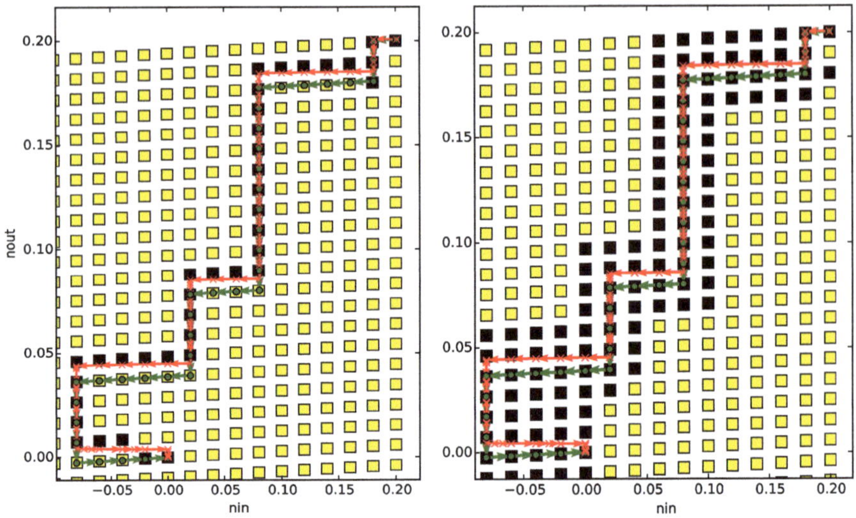

Fig. 1.10 Different methods for selecting the *covered* states of a simulation result. *Green* points indicating wanted points with a path finding algorithm (described more in detail in Sect. 1.5.3), *red* crosses the corresponding correct transient simulation result points, *black* boxes are marked as covered using nearest neighbor (*left*) or Euclidean distance (*right*)

is the set of states in Σ_R which were covered by the simulation response S. Since every point s_i has a nearest neighbor, the distance is not considered (cf. Fig. 1.10). Hence, if a discretization only consists of very few states, each point of a simulation response will lead to a covered state, although the state is very far off. Obviously, this simple approach does not calculate a smooth and adequate measure.

A better approach would be to use an Euclidean distance to cover all states in a given region around the transient simulation result (cf. Fig. 1.10), eliminating the nearest neighbor problems above mentioned. Additionally, it will allow having a measure independent of the sampling distance in the state space as well as the sampling distance of the transient simulation result.

However, a maximum distance must be chosen adequately, since using a too large distance could mark states with different behavior compared to the transient trajectory under investigation, while a too small distance will underestimate the set of covered states. A good starting point for the distance is to select the *median* distance between two neighbor states in the discrete state space or to use a percentage of the diameter of the reachable state space. Here, we conservatively take the median length of all transitions R inside the ATS.

Consequently, we are now able to compute the coverage using the cardinality of the elements in the set C and the number of states in the reachable discrete state space Σ_R:

$$\zeta = \frac{|C|}{|\Sigma_R|} \tag{1.10}$$

Equation (1.10) indicates two ways of enhancing the coverage ζ: increasing the number of covered states $|C|$ by running simulations until a desired coverage measure is reached or decreasing the number of states $|\Sigma_R|$ by analyzing the whole analog state space more in detail (as described in Sect. 1.6).

1.5.2 Coverage Maximization Algorithm

Since one single transient simulation response covers only a small amount of states, we introduce an algorithm to cover all reachable states of the ATS. For this, we enhance the ATS with a labeling function $\omega : \Sigma \rightarrow \mathbb{N}_0^+$ that labels each state with a weight, indicating how often this state was already visited by the algorithm. Combined with the total transition relation R, this information eases the path finding algorithms which will be introduced subsequently.

In each step, the algorithm selects a state with minimum weight and calculates a path using an A^* search. Selecting longest paths with minimum cost maximizes the possibility to cover the most unvisited points at once. While traversing the graph, we are able to create an input stimulus for a simulation. The resulting transient response is now used to calculate a coverage value ζ for this single stimulus. In the last step, the weights of the covered states are updated for the next iteration step. By increasing the node weight every time a state is covered by a simulation response, this node

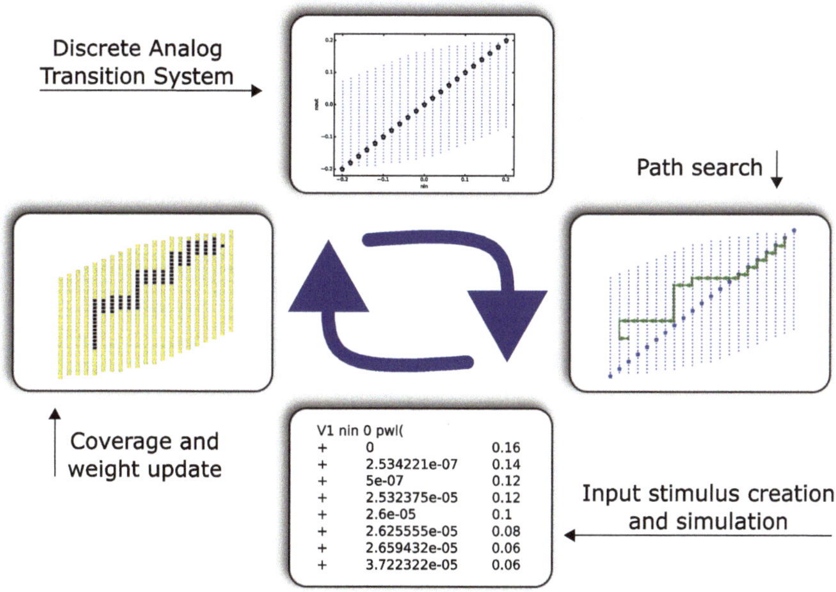

Fig. 1.11 Coverage maximization algorithm based on discrete state space modeling

is avoided in the future path finding attempt. The coverage maximization algorithm using information from the discrete state space model is illustrated in Fig. 1.11.

1.5.3 Path Planning

The creation of appropriate input stimuli is crucial for the method described in the previous section. By visiting each state in the ATS in a single stimulus, a vast number of very small simulations is needed. As a result, the startup time of the simulation software will dominate the simulation time. Consequently, a path planning algorithm is needed to create simulation input stimuli which meet the following characteristics:

- The resulting path should avoid already visited states.
- It should consist of as many unvisited states in the ATS as possible.
- The length of the path regarding covered states should be maximized with respect to the criteria described before.

An approach to satisfy these criteria exists in [21], where a stimulus is created by traversing the whole graph with one single path. With larger circuit size (resulting in more state-space dimensions and more state-space points), however, a full input stimulus created using this method consists of significantly more points than the ATS itself, thus resulting in a very long runtime of the simulation. The constructed single stimulus by that method additionally performs badly in terms of the achieved state space coverage as we can see in the small low-pass example at the end of this section.

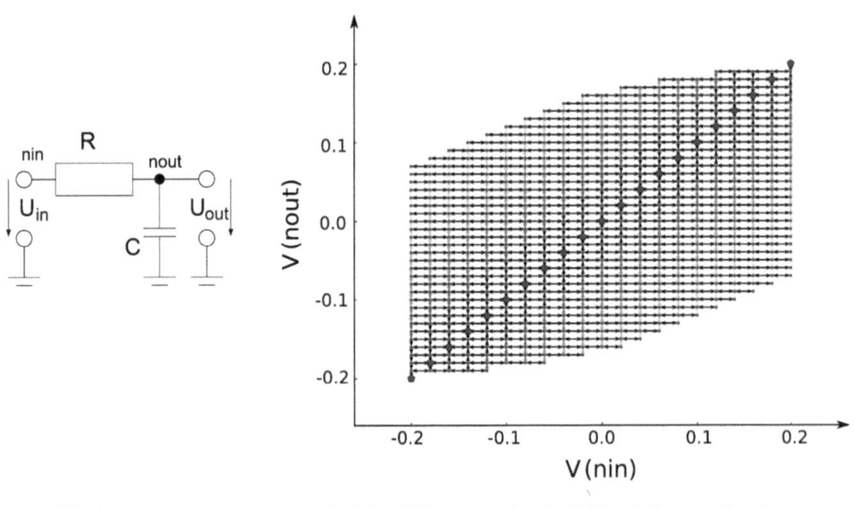

Fig. 1.12 State space coverage example: Simple low-pass circuit (*left*) and the resulting ATS system after a reachability analysis

Table 1.1 Results of the presented path planning algorithm. The sum of data points is the length of all created input stimuli for the simulation

Path planning method	Number of simulations	Sum of data points	Resulting coverage (%)	Runtime
Presented	4	895	100.00	5.8 s
Single	1	3213	97.33	12.1 s

We now introduce a weight-based path planning algorithm. Let π be a set of states describing the path from a starting to a target state. The length $|\pi|$ is the number of states inside the path. Since every state has a weight ω_σ, the weight of a path is $\omega_\pi = \Sigma_i^n \omega_{\sigma_i}$. Additionally, the set Σ_{DC} is a set of DC operation points of the ATS.

We now randomly choose an unvisited state σ_u, indicated by its weight $\omega_\sigma = 0$ and compute iteratively *all* paths, beginning in the set of DC operating points. To increase the number of visited states, additionally all paths from the target state back to a DC operating point are searched. If a path exists, we can start again to another randomly chosen point, allowing the possibility of concatenating the resulting paths. To avoid very long paths, the length of the resulting path is limited by the number of unvisited states in the whole system.

The results of the path planning algorithm described in this section can be found in Table 1.1. The algorithm was used with an ATS created from a simple low-pass circuit (see Fig. 1.12), which consists of 650 states. Even with this very basic and simple analog circuit, the number of data points of the generated simulator stimuli and the simulation time are much lower as for the former *single* method. As we will see later on, the speedup is much bigger in analog circuits with higher complexity.

Since this algorithm tries to reach every single state in the state space and due to the limited path length, the total number of simulations, and the resulting runtime are determined by the size of the inspected reachable state set. With increasing complexity of the investigated circuits, it is furthermore not possible to reach high coverage measures ζ in a reasonable computation time and with short overall input stimuli length. Hence, the number of states to inspect must be reduced, which will be described in the next section.

1.6 λ State-Space Coverage

As mentioned in the previous section, trying to visit every single reachable state in an analog circuit might not make sense. To reduce the number of states to cover, while maintaining a high probability to find all bugs—which translates in visiting only important states—we have to utilize further system characteristics represented in the ATS. Therefore, we are segmenting the discrete state space to regions with uniform (linear) behavior which only needs one simulation trajectory through this region to

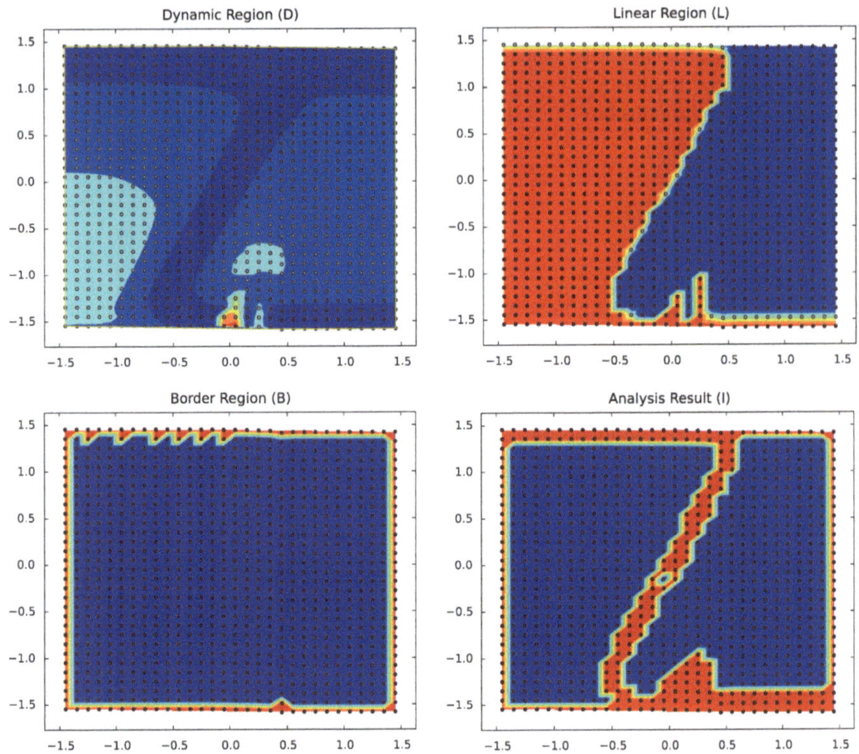

Fig. 1.13 State-space analysis steps, from *top left* to *bottom right*: Dynamic region (D) detection, linear region (L) detection, border region (B) detection, interesting region (I) detection results. The last diagram (I) visualizes the resulting interesting states (*red color* $w_\sigma = 0$)

ensure the correctness of the implementation. By contrast, as analog systems also exhibit nonlinear parts with high dynamic (such as limited output voltage swings), these regions need to be thoroughly visited under all circumstances as the probability for faulty behavior in these nonlinear regions is much higher than in uniform linear regions.

Hence, in these regions, every state should be covered to detect possible problems in the implementation. For distinguishing uniform linear from nonlinear nonuniform regions, the eigenvalues captured in the states of the ATS are used. Regions with numerically similar eigenvalues are detected and marked as uniform. Additionally, border regions and the direct neighborhood of a DC operation point have to be covered, because the probability of errors at borders is high and DC points are mandatory to visit.

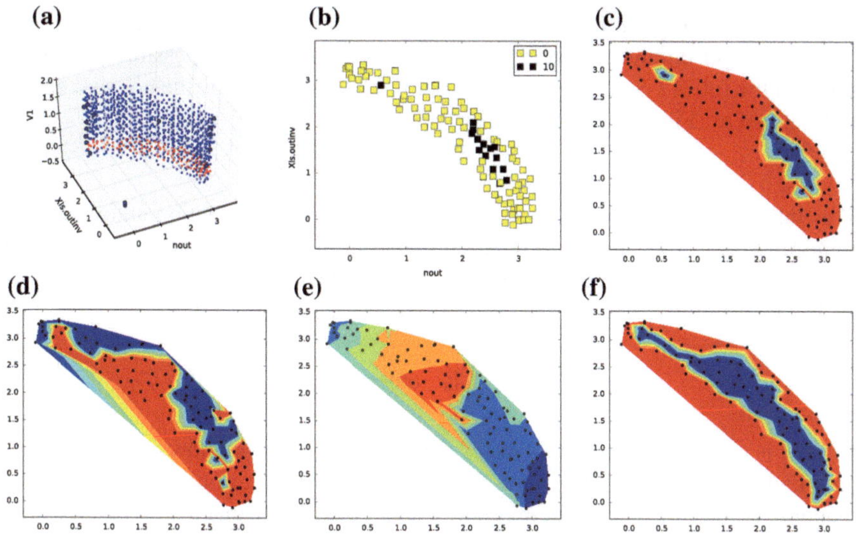

Fig. 1.14 Results of the λ state-space analysis of the level-shifter circuit: **a** shows the reachable set in the state space. The *yellow* points in (**b**) show interesting states to be covered (i.e. the Σ_λ set), indicated by $w_\sigma = 0$ based on the interest value I_σ shown in (**c**). *Red* in (**d**) marks areas with high dynamic, based on the eigenvalue in (**e**). **f** shows border regions. For visualization purposes, (**b**–**f**) are plotted for the plane $V1 = 0.0V$

To compute the λ state-space coverage, four different coverage value vectors have to be computed:

- Linear regions of the system are detected using the eigenvalues. Each state σ in the system has a list of ancestors and successor states. L_σ is set to 1 if one of the neighboring states has a significantly ($|\cdot| > 50\%$) different eigenvalue than σ, otherwise it is set to 0.
- The median of all eigenvalues of the whole ATS is computed, indicating the basic dynamic level of the analog circuit. D_σ is then the absolute difference to the median value for each state $\sigma \in \Sigma_R$. All values are normalized to [0..1].
- To compute the states in the border region of the reachable set, the convex hull $conv(\Sigma_R)$ of all reachable states of the circuit is computed using the approach from [45]. B_σ is set to 1 if the state σ is located on the edges of the resulting polytope, otherwise it is set to 0.
- For each DC operating point, the direct neighborhood is computed, using the k-d tree described before. O_σ is set to 1 if the state σ lies in the neighborhood of the DC operating point or is the state itself.

For illustration, Fig. 1.13 visualizes the result of the state-space analysis for a Schmitt trigger circuit. As every step of the analysis indicates possible interesting states of the full ATS system, the region of interest is formed by the nonzero entries in $\mathbf{I}$ defined as (Fig. 1.14):

$$\mathbf{I} = \mathbf{L} + \mathbf{D} + \mathbf{B} + \mathbf{O} \tag{1.11}$$

This information can now be integrated in the coverage calculation algorithm presented in Sect. 1.5. For this purpose, each state is initialized with a node weight which is determined by the corresponding region of interest factor. The initial weight for each state is defined as

$$w_\sigma = \begin{cases} 0, & \text{if } I_\sigma \geq t, \\ 1, & \text{otherwise.} \end{cases} \tag{1.12}$$

where t is a given threshold. All states with a low weight[1] are now preferred by the path finding algorithm. States with a higher weight are not removed from the path planning algorithm, so that there is still a small possibility that a simulation covers this state. With these weights, we can define a reduced set Σ_λ of interesting state space points based on the reachable set Σ_R:

$$\Sigma_\lambda = \{\sigma \in \Sigma_R | w_\sigma = 0\} \tag{1.13}$$

The λ *state space coverage* can now be defined as the number of visited states divided by the number of states in the reduced set Σ_λ, where in the numerator only states are counted which belong to that reduced set Σ_λ:

$$\zeta_\lambda = \frac{|\{\sigma \in C | \sigma \in \Sigma_\lambda\}|}{|\Sigma_\lambda|} \tag{1.14}$$

1.7 Coverage Analysis and Optimization Results

For the experiments, we have used 6 analog transistor-level circuits:

- A simple RC low-pass filter,
- a Schmitt trigger circuit,
- an active band-pass circuit (see Fig. 1.16 for a schematic),
- a level-shifter circuit,
- a log domain filter,
- and an industrial gm-C low-pass filter (see Fig. 1.15 for a schematic).

[1] A low weight ($w_\sigma = 0$) indicates an interesting state inside the ATS. The weight is set to zero, because this can be directly be used as the initial value for the labeling function ω, which indicates how often this state was covered before or not, which is used by the path planning algorithm presented in Sect. 1.5.3.

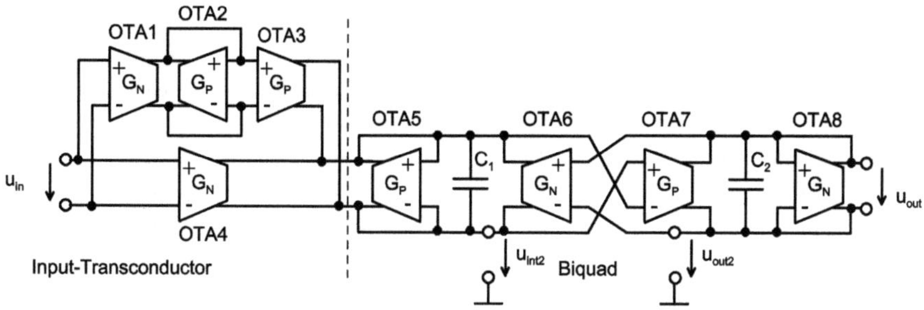

Fig. 1.15 Schematic of the gm-C filter circuit

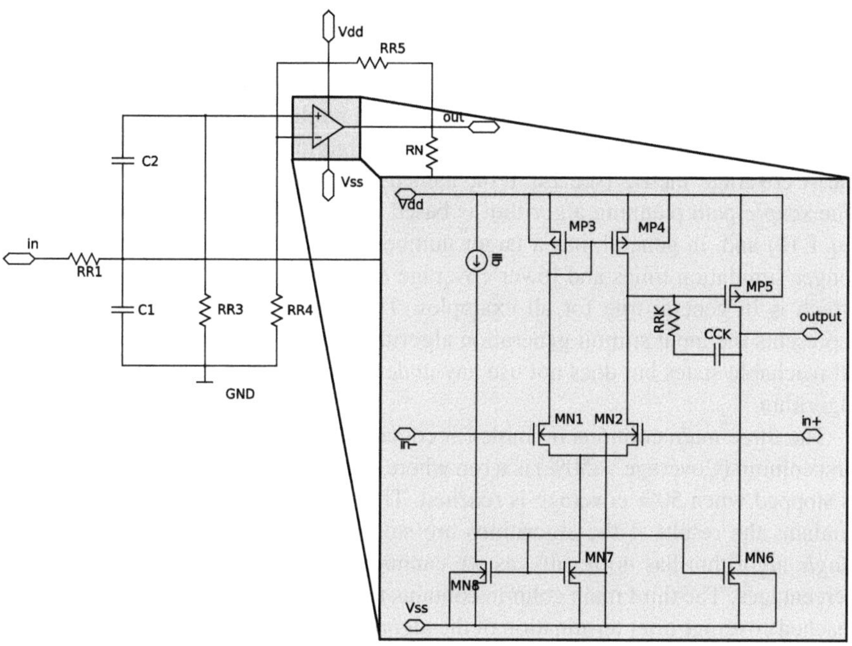

Fig. 1.16 Schematic of the active RC band-pass circuit using a miller-compensated operational amplifier

Some circuit statistics and discretization results of the circuits are shown in Table 1.2. All used transistors are MOS-transistors from a 180 nm PTM technology and are simulated/verified with a full BSIM3v3-model and full accuracy. The experiments are carried out on a 3.4 GHz Dual-Core machine. The simulations of the input stimuli for all circuits except the gm-C filter are performed with the SPICE-simulator GNUCAP. The gm-C filter is discretized and simulated with an industrial in-house simulator which is somewhat faster. However, due to the full SPICE-accuracy the results are very accurate. The third and the fourth column of Table 1.2 gives the number of states in the discretized reachable set and reduced reachable set, respectively.

Table 1.2 Statistics of applied analog circuits on transistor level

Analog circuit	Transistors	State space dimensions	States in reachable set $\|\Sigma_R\|$	States in λ reduced set $\|\Sigma_\lambda\|$
RC low-pass filter	0	2	861	227
Schmitt trigger	10	2	1903	356
Band-pass filter	8	3	5660	1948
Level shifter	6	3	1081	641
log domain filter	13	2	2526	394
gm-C filter	69	3	344	239

One can notice, that for the linear circuits (RC low-pass, band-pass), the reduction factor is—as expected—much larger than for the nonlinear circuits.

In Table 1.3, we compare the path-planning algorithm from Sect. 1.5.3 based on the λ coverage metric (see Eq. 1.14, named λ-*Cov.*) with two other approaches. The *simple* path planning algorithm is based on the standard coverage metric (see Eq. 1.10) and, in general, has a larger number of states to visit. Hence, we expect longer simulation times and lower coverage compared to the λ-*Coverage* method, which is in general true for all examples. The third method is called *single* and represents the input stimuli generation algorithm from [21] which tries also to visit all reachable states but does not use any underlying coverage guided path-planning algorithm.

The three main columns in Table 1.3 contain the results for three test cases: the first column (Coverage > 50%) is a run where the generation of further input stimuli is stopped when 50% coverage is reached. The second column (Coverage > 75%) contains the results if the algorithms are stopped at 75%. For both columns, the *single* algorithm has no results, as we cannot interrupt the algorithm at the given percentages. The third main column contains the results for a full run, including the reached coverage after termination of the algorithms and the number of input stimuli (where each stimulus is run in a single simulation) to get there and finally the runtime of the algorithm in seconds.

In all three columns and for all circuits, the λ-*Coverage* methodology has the shortest runtimes and highest coverage. On average, the λ-*Coverage* method is 5.8 times faster than the *simple* methodology and 5.4 times faster than the *single* method. The increase in coverage is on average 7.9% compared to the *simple* and 18.8% compared to the *single* method.

For bigger circuits, no algorithm reaches 100% coverage due to the over-approximating reachable set and due to discretization errors. In other words, there exist some points in the set Σ_R and also in Σ_λ which are assumed to be reachable. However, if one tries to reach that state with a calculated input stimulus, the circuit will eventually not be able to get into that state. We use a randomized correction algorithm to finally

Table 1.3 Results of the proposed coverage calculation algorithm. *simple* is the described path-planning algorithm on the full state space, λ-*Cov.* is the proposed path-planning algorithm with underlying λ state-space coverage metric, and *single* is the path planning algorithm from [21]

Analog Circuit	Method	Coverage > 50%		Coverage > 75%		Overall Coverage in %		
		# Sim.	Runtime	# Sim.	Runtime	Coverage	# Sim.	Runtime
Schmitt trigger	*simple*	3	98.96	5	134.06	ς 87.06	17	215.83
	λ-*Cov.*	1	5.62	2	8.16	ς_λ 91.40	3	10.79
	single	1	19.29	1	19.29	ς 30.32	1	19.29
Band-pass filter	*simple*	6	74.87	15	124.95	ς 83.06	173	556.86
	λ-*Cov.*	1	31.08	6	97.37	ς_λ 88.99	18	164.75
	single	1	388.96	–	–	ς 62.59	1	388.96
Level shifter	*simple*	–	–	–	–	ς 43.91	9	81.46
	λ-*Cov.*	1	12.81	–	–	ς_λ 72.21	6	27.97
	single	1	266.24	–	–	ς 66.39	1	266.24
log domain filter	*simple*	1	1.91	2	2.15	ς 100.00	7	3.23
	λ-*Cov.*	1	0.40	1	0.40	ς_λ 100.00	4	1.66
	single	1	16.65	1	16.65	ς 100.00	1	16.65
gm-C filter	*simple*	2	0.58	7	2.98	ς 76.57	8	3.98
	λ-*Cov.*	1	0.47	3	1.18	ς_λ 85.17	4	1.92
	single	1	8.59	–	–	ς 65.71	1	8.59

hit that point. After some tries this algorithm will stop, leaving behind some states as unvisited. Hence, the given coverages in Table 1.3 would probably be closer to 100%, if we knew the exact reachable set and if we were able to find all exact stimuli to all states in that set.

1.7.1 Detailed Case Study of a Level-Shifter Circuit

For further insights, we now elaborate on the results from the level-shifter. In Fig. 1.14a, we see the discretized reachable state space spanned by the input $V1$ and the state variables $nout$ and $Xls.outinv$. The red plane is selected and shown in the other five 2-D plots. Figure 1.14e shows the absolute value of the first eigenvalue. Calculated from these eigenvalues, Fig. 1.14d displays the areas with dynamic behavior resulting in **D** (see Sec. 1.5). Combining Fig. 1.14d and the states to visit from

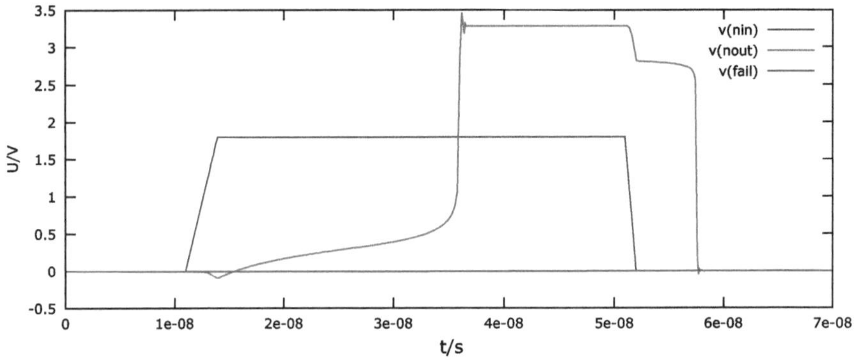

Fig. 1.17 Simulation of the level-shifter circuit using a manually generated straightforward input stimulus $v(nin)$. $v(fail)$ is high, if the logic output level ($>80\%=1$, $<20\%=0$) does not coincide with the logic input level for more than 50 ns. Here, no fail signal appears

the border calculation in Fig. 1.14f results in the overall selection function according to Eq. (1.12) in Fig. 1.14c. Mapping that back to a discrete decision, we end in the "to be covered" set Σ_λ marked yellow in Fig. 1.14b. For all other $V1$ planes in the 3-D state space similar pictures exist, resulting in an overall "to be covered" set Σ_λ, concentrating on the nonlinear and border regions of the level-shifter circuit. With this information, input stimuli with a total λ-Coverage of ζ_λ = 72.21% could be generated by the proposed algorithm.

Assume now that we test only with some straightforward input stimuli such as the one shown in Fig. 1.17 as $v(nin)$. For a simple and reliable investigation we attached a monitor circuit to the level shifter, which sends a fail signal if the input and the output of the level shifter stay at opposite logic values for more than 50 ns.

The zero fail signal suggests the correct function of the circuit. Manual inspection will also suggest that the delay is in expected bounds and that the logic levels are reached in a proper manner. A designer could conclude that the circuit has a delay of about 25ns and will function correctly. However, if the described λ-Coverage metric is used, we will first see, that the used input stimulus of Fig. 1.17 only has $\zeta_\lambda = 13.9\%$ coverage, and secondly that the generated stimuli with $\zeta_\lambda = 72.21\%$ coverage will uncover a bug shown in Fig. 1.18. The latter figure shows the results from a generated stimulus where the fail signal does not disappear for more than 50ns and furthermore will keep the level-shifter in a faulty state. Consequently, this example shows that usage of the λ state space coverage method can help to systematically find unknown bugs in analog circuits.

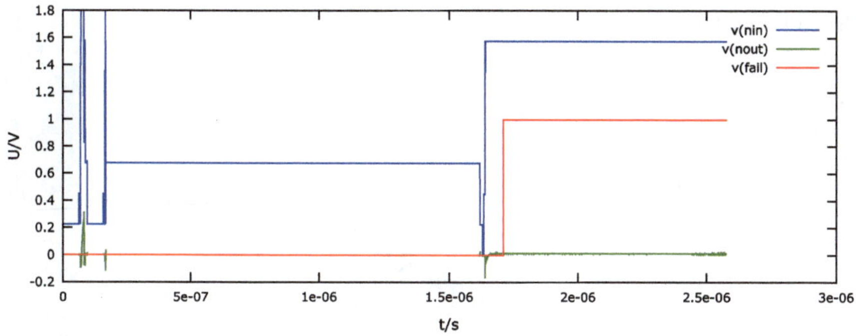

Fig. 1.18 Generated input stimulus using the λ state coverage path finding algorithm showing a lasting fail state, which is a bug in the implementation of the level-shifter, identified by the proposed approach

1.8 System-Level Verification

In order to bring formal verification of analog circuits and systems up to system level, we have some options:

- Use hybrid system level verification tools to analyze and verify the system-level models.
- Use analog verification tools with abstracted behavioral models. Then partly replace the behavioral models or equations with circuit implementations and repeat trying to check the correctness.

However, the latter approach will stop at a certain abstraction level due to complexity issues. The first approach is even not able to go down the hierarchy. To further close the gap between system level and transistor level we can use the following strategies.

- Compare the models on system level using equivalence checking with the lower level circuits and systems.
- Use coverage analysis to generate stimuli on lower level to also compare behavioral models to the circuit level. Use these stimuli on any level to check correctness.

Here, we want to give an instructive example for the system level and explain some of the options mentioned above. We use a model of a robot which follows a line on the ground (see Fig. 1.19). The most compact abstraction results in three explicit ordinary differential equations (ODEs) which can be expressed in MATLAB, Space-Ex, or VerilogA:

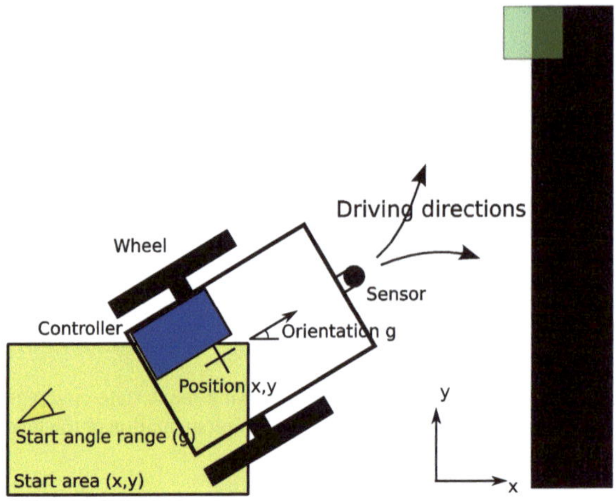

Fig. 1.19 Line following robot. Depending on the x, y, start position and the orientation g (ϕ), the controller is able to follow the *left* edge of the *black* half plane on the floor using the light sensor in front of the robot

$$\begin{aligned}
\dot{x} &= v_0 \cdot cos(\phi) \\
\dot{y} &= v_0 \cdot sin(\phi) \\
\dot{\phi} &= 0.5 \cdot (tanh\,(10 \cdot (L \cdot cos(\phi) + x - k)) + 1) \cdot \frac{w \cdot v_0}{\pi \cdot B} + \\
&\quad 0.1 \cdot \frac{-1 \cdot w \cdot v_0}{\pi \cdot B}
\end{aligned} \tag{1.15}$$

with $L = 0.1\,m$, the distance of the sensor to the center of the robot, $B = 0.05\,m$, the width of the robot, $v_0 = 0.5\,\frac{m}{s}$, the speed of the robot, $k = 0.2\,m$, the position of the black half plane on the ground, $w = 0.5$, the ratio between speed and angular speed, x and y, the position of the robot and ϕ, the driving direction of the robot. In the following, the driving direction ϕ will be named g for ease of writing in plots and specifications. The last equation of Eq. (1.15) models a two point switching controller with a smooth transition between the states.

As theses ODEs are nonlinear we cannot use tools like Space-Ex [16] or the reachability based on Minkowski sums [46] directly. However, building a piecewise-linear hybrid model would enable the use of both approaches and would lead to a nice reachability analysis result. In our case, we can formulate the problem of robots' locking to the line and sticking on that line as a reachability problem using a start point and direction as an initial condition and checking if that trajectory reaches a target area (see green box in Fig. 1.19). If we use a set of starting points and directions (illustrated with the yellow box in Fig. 1.19), we calculate the ATS in 867 s on a 2.27 GHz Intel Xeon Machine and prove that the specification will hold on all start conditions in the given start area using the ASL description shown in Fig. 1.20.

In essence, the description demands, that all trajectories coming from the start area finally end in the goal area. In Fig. 1.21(left), the positive result of the proof is illustrated with all trajectories starting in the start area and ending inside the target area. Both areas are separately shown in Fig. 1.21(right). Additionally, the stability

```
target = 'V(NY)'[>3.8] and 'V(NX)'[>-0.15] and 'V(NX)'[<0.16] and
         'V(NG)'[>1.3] and 'V(NG)'[<1.9] ;

start_area = 'V(NY)'[>-0.1] and 'V(NY)'[<0.1] and
             'V(NX)'[>-0.25] and 'V(NX)'[<0.17] and
             'V(NG)'[>1.3] and 'V(NG)'[<1.95] ;

reachable = EF^-1 start_area ;

fail = reachable and 'V(NY)'[>1.95]  and ( not target ) ;

for all assert (!fail);       # fail should be empty
```

Fig. 1.20 ASL property for a stable run of the robot

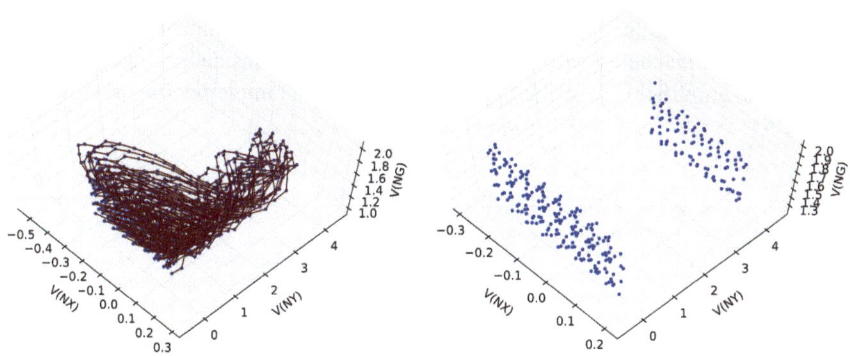

Fig. 1.21 *Left* Trajectories starting in the start area for the line follower on top level. *Right* Start area (*front*) and target area (*back*) in the same state space

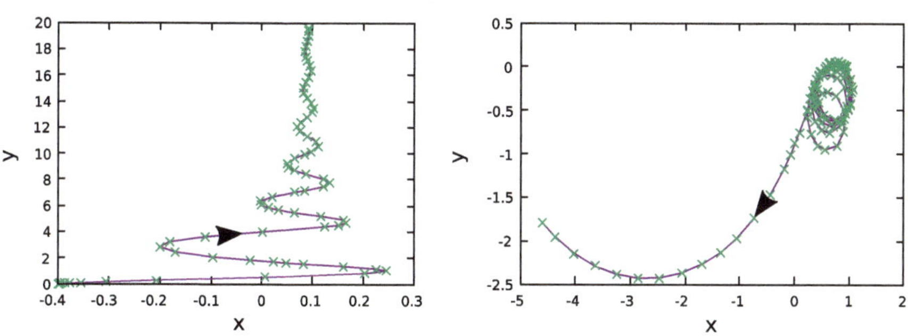

Fig. 1.22 Trajectories from simulations of the line follower: *Left* Correct motion along the y-axes around $x = k$. *Right* Wrong motion due to wrong start point and direction

of the behavior in that region is indirectly also proved, as the start area is slightly larger and includes the target area in terms of the position x and the driving angle ϕ. Hence, we can conclude that the system will probably evolve by contracting and not expanding.

In Fig. 1.22, two simulations are shown in the x-y plane to illustrate the correct and faulty motion of the robot. If it is started to far away from the line, it cannot lock to that line.

1.8.1 System Refinement and Verification

Finally, we want to show how a closed chain of proof from system level down to transistor level could look like. In order to refine the system and take real implementations of parts of the system into account, we concentrate on the controller (see Fig. 1.23 for a complete overview of the hierarchical decomposition of the system and the formal verifications performed).

First, the controller is replaced by a behavioral description of a decision unit and a netlist of an analog buffer amplifier (see Fig. 1.23). Inside the analog buffer an operational amplifier is described by a behavioral model implementing the following behavior:

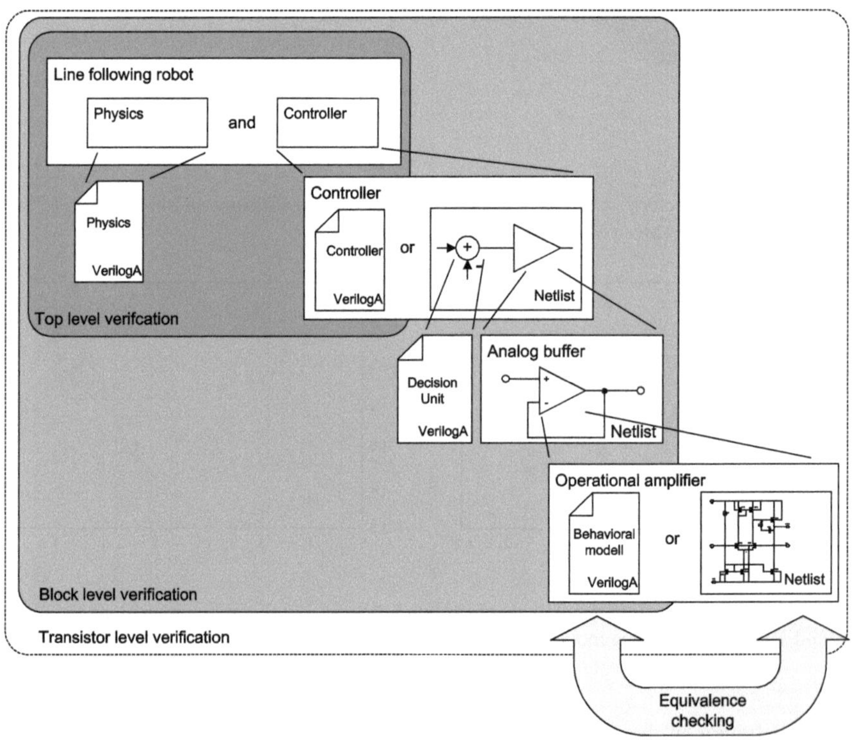

Fig. 1.23 Hierarchical decomposition of the line following robot down to transistor level and verification steps performed

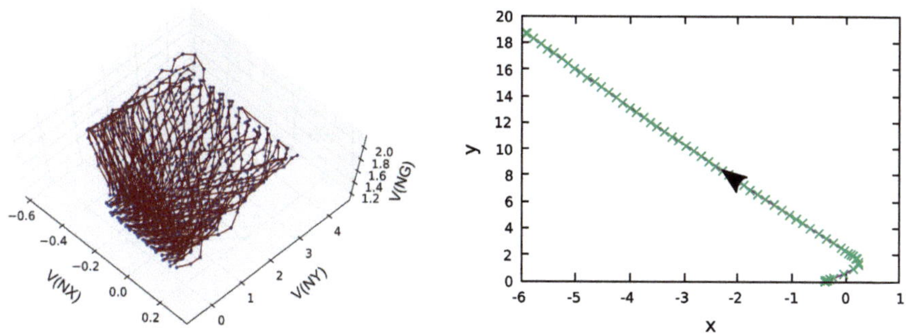

Fig. 1.24 *Left* Trajectories of the reachable area for the line follower with a controller using a behavioral description with too small output swing resulting in failing trajectories clearly not hitting the target area. *Right* Motion of the robot following one of these failing trajectories

- One pole for the dynamics,
- limiting of output voltage to supply rails,
- limiting of input voltage to an input voltage range and
- maximal slew rate of output.

Using these models increases the runtime to 3079 seconds of the block-level verification. However, the proof of the property also succeeds.

If we insert a behavioral model of the amplifier with tight output limitation ($\pm 0.4V$), a fail will be the result of the model check. A trajectory which is not damped down to follow the line is given as a counterexample. See Fig. 1.24(left) for all failing trajectories in the state space and Fig. 1.24(right) for one simulation of a failing one in the x-y plane.

Next step is a proof that the operational amplifier (OP) is equivalent to a transistor implementation of the OP. We use an OP with the structure of Fig. 1.16. The equivalence checking with a given maximal slew rate on the input of the amplifier of $5\frac{V}{\mu s}$ succeeded with an error ϵ_{dyn} of below 10% (see Fig. 1.25 for a plot of the errors over the reachable area).

In practice, the formal proof chain is now closed from transistor to system level. However, as we are—in this small example—able to replace the behavioral model of the amplifier by its transistor implementation, a direct proof of the transistor-level implementation has been carried out. It takes 5752 seconds and succeeds also to prove the wanted properties of the system level. Furthermore, the last equivalence-checking result could also be achieved if we use the proposed coverage metric to generate stimuli for checking the netlist of the amplifier versus its behavioral model. Taking all together, we can conclude that the line following robot will expose the wanted functionality also with the full detailed implementation.

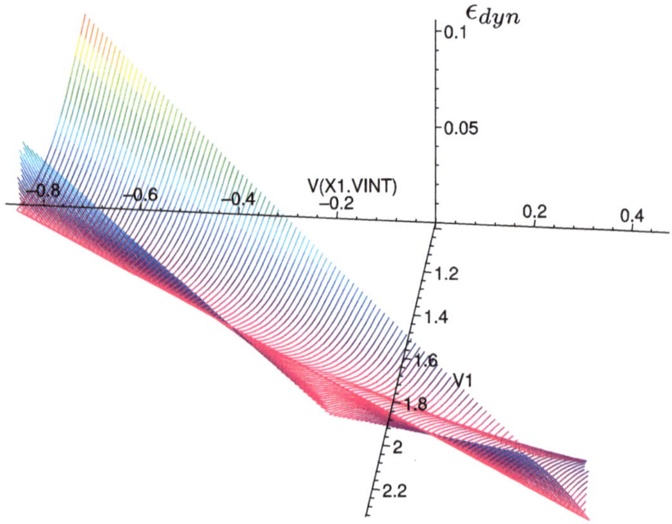

Fig. 1.25 Dynamic error ϵ_{dyn} of the equivalence check of the behavioral model of the amplifier and the transistor netlist

A missing extension in this context is the inclusion of digital circuitry on system level. In that case, for small digital parts the mentioned hybrid system verification tools can be used or digital controllers may be continuously abstracted. However, for big mixed-signal system-level or block-level descriptions powerful tools for formal verification are still missing.

1.9 Conclusion

In this chapter an overview of formal verification techniques ranging from transistor level to system level for analog circuits and systems has been presented. Main methods are reachability analysis in different flavors, model-checking methods and equivalence-checking methods. The focus in this chapter lays on sampling methods allowing an accurate solution of the underlying DAE system. Actual improvements allow the calculation of accurate and efficient coverage measures to improve the individual validation of blocks. Furthermore, the methods are extended to system level allowing a hierarchical formal verification flow closing a chain of proof from transistor level up to system level.

References

1. A. Vachoux, C. Grimm, K. Einwich, A. Vachoux, C. Grimm, K. Einwich, Analog and mixed-signal modeling with SystemC-AMS, in *Proceedings of the 2003 International Symposium on Circuits and Systems, 2003, ISCAS'03*, vol. 3 (2003), pp. III–914
2. R. Telichevesky, K. Kundert, J. White, SpectreRF; New Frontiers in Circuit Simulation (1995), pp. 3–24
3. D. Zaum, S. Hoelldampf, M. Olbrich, E. Barke, I. Neumann, S. Schmidt, The PRAISE approach for accelerated transient analysis applied to wire models, in *Behavioral Modeling and Simulation Workshop, BMAS 2009* (IEEE, 2009), pp. 120–125
4. G. Gielen, E. Maricau, in *Analog IC Reliability in Nanometer CMOS* (Springer, Heidelberg, 2013)
5. T. Henzinger, P.-H. Ho, H. Wong-Toi, HyTech: a model checker for hybrid systems. Lecture Notes in Computer Science (Springer, 1997), pp. 460–463
6. E. Asarin, T. Dang, O. Maler, d/dt: a tool for reachability analysis of continuous and hybrid systems (2001)
7. A. Eggers, N. Ramdani, N.S. Nedialkov, M. Fränzle, Improving the SAT modulo ODE approach to hybrid systems analysis by combining different enclosure methods. Softw. Syst. Model. **14**(1), 121–148 (2012)
8. T.A. Henzinger, P.-H. Ho, H. Wong-Toi, Algorithmic analysis of nonlinear hybrid systems. IEEE Trans. Autom. Control **43**(4), 540–554 (1998)
9. G. Frehse, PHAVer: algorithmic verification of hybrid systems past HyTech, in *Proceedings of Hybrid Systems: Computation and Control (HSCC 2005)*. Lecture Notes in Computer Science, vol. 3414 (2005), pp. 258–273
10. T. Dang, A. Donzé, O. Maler, Verification of analog and mixed-signal circuits using hybrid system techniques, in *Formal Methods in Computer-Aided Design* (Springer, 2004), pp. 21–36
11. Z.J.D. Ghiath Al Sammane, Mohamed H. Zaki, S. Tahar, Towards assertion based verification of analog and mixed signal designs using PSL, in *Forum on Design Languages (FDL)* (2007)
12. S. Little, D. Walter, N. Seegmiller, C. Myers, T. Yoneda, Verication of analog and mixed-signal circuits using timed hybrid petri nets, in *ATVA* (2004), pp. 426–440
13. L. Hedrich, E. Barke, A formal approach to verification of linear analog circuits with parameter tolerances, in *Proceedings of the Design, Automation and Test in Europe* (1998), pp. 649–654
14. D. Grabowski, M. Olbrich, C. Grimm, E. Barke, Analog circuit simulation using range arithmetics, in *ASPDAC* (2008), pp. 762–767
15. M. Althoff, A. Rajhans, B.H. Krogh, S. Yaldiz, X. Li, L. Pileggi, Formal verification of phase-locked loops using reachability analysis and continuization. Commun. ACM **56**(10), 97–104 (2013)
16. G. Frehse, C. Le Guernic, A. Donzé, S. Cotton, R. Ray, O. Lebeltel, R. Ripado, A. Girard, T. Dang, O. Maler, SpaceEx: scalable verification of hybrid systems. in *Computer Aided Verification* (Springer, 2011), pp. 379–395
17. M. Freisfeld, M. Olbrich, E. Barke, Circuit simulations with uncertainties using affine arithmetic and piecewise affine statemodels, in *9th International Conference on Solid-State and Integrated-Circuit Technology, 2008 ICSICT* (2008), pp. 496–499
18. T.R. Dastidar, P.P. Chakrabarti, A verification system for transient response of analog circuits. ACM Trans. Des. Autom. Electron. Syst. **12**(3), 31:1–31:39 (2008)
19. L. Hedrich, E. Barke, A formal approach to nonlinear analog circuit verification, in *Proceedings of the IEEE/ACM International Conference on Computer-Aided Design ICCAD-95*. Digest of Technical Papers (1995), pp. 123–127
20. A.V. Karthik, S. Ray, P. Nuzzo, A. Mishchenko, R. K. Brayton, J. Roychowdhury, ABCD-NL: approximating continuous non-linear dynamical systems using purely Boolean models for analog/mixed-signal verification. in *ASP-DAC* (2004), pp. 250–255
21. S. Steinhorst, L. Hedrich, Improving verification coverage of analog circuit blocks by state space-guided transient simulation, in *IEEE International Symposium on Circuits and Systems* (2010)

22. D. Grabowski, D. Platte, L. Hedrich, E.A. Barke, Time constrained verification of analog circuits using model-checking algorithms, in *ENCTS: Workshop on Formal Verification of Analog Circuits* (2005)
23. O. Maler, A. Pnueli, Extending PSL for Analog Circuits, in *Research Report:PROSYD: Property-Based System Design FP6-IST-507219* (2005)
24. S. Steinhorst, L. Hedrich, Analog assertion-based verification on partial state space representations using ASL. In *2012 Forum on Specification and Design Languages (FDL)* (2012), pp. 98–104
25. J. Eckmüller, M. Gröpl, H. Gräb, Hierarchical characterization of analog integrated circuits, in *DATE '98: Design, Automation and Test in Europe* (1998)
26. M. Ma, L. Hedrich, C. Sporrer, ASDeX: a formal specification for analog circuit enabling a full automated design validation. in *Design Automation for Embedded Systems* (2012), pp. 1–20
27. S. Fine, A. Ziv, Coverage directed test generation for functional verification using Bayesian networks, in *Proceedings of the Design Automation Conference* (2003), pp. 286–291
28. F. Haedicke, D. Große, R. Drechsler, A guiding coverage metric for formal verification, in *Design, Automation and Test in Europe Conference and Exhibition (DATE)* (IEEE, 2012), pp. 617–622
29. J.-Y. Jou, C. Liu, Coverage analysis techniques for hdl design validation. in *Proceedings of the Asia Pacific CHip Design Languages* (1999), pp. 48–55
30. K. Arabi, B. Kaminska, Parametric and catastrophic fault coverage of analog circuits in oscillation-test methodology, in *15th IEEE VLSI Test Symposium* (1997), pp. 166–171
31. J. Parky, S. Madhavapeddiz, A. Paglieri, C. Barrz, J. Abraham, Defect-based analog fault coverage analysis using mixed-mode fault simulation. in *IEEE 15th International Mixed-Signals, Sensors, and Systems Test Workshop, 2009, IMS3TW '09* (2009), pp. 1–6
32. M. Horowitz, M. Jeeradit, F. Lau, S. Liao, B. Lim, J. Mao, Fortifying analog models with equivalence checking and coverage analysis, in *Proceedings of the 47th Design Automation Conference, DAC '10, NY, USA, New York* (2010), pp. 425–430
33. A. Julius, G. Fainekos, M. Anand, I. Lee, G. Pappas, Robust test generation and coverage for hybrid systems, in *Proceedings of the 10th International Conference on Hybrid Systems: Computation and Control (HSCC)* (2007), pp. 329–342
34. T. Nahhal, T. Dang, Test coverage for continuous and hybrid systems. in *Computer Aided Verification* (Springer, 2007), pp. 449–462
35. L. Pillage, C. Wolff, R. Rohrer, Dominant Pole(s)/Zero(s) analysis for analog circuit, in *CICC* (1989), pp. 21.3.1–21.3.4
36. R. Freund, P. Feldmann, Reduction-order modeling of large linear passive multi-terminal circuits using matrix-pade approximation, in *Review* (1988), pp. 1–19
37. S. Steinhorst, L. Hedrich, Advanced methods for equivalence checking of analog circuits with strong nonlinearities. Form. Methods Syst. Des. **36**(2), 131–147 (2010)
38. S. Steinhorst, L. Hedrich, Trajectory-directed discrete state space modeling for formal verification of nonlinear analog circuits, in *Proceedings of the International Conference on Computer-Aided Design* (ACM, 2012), pp. 202–209
39. W. Hartong, R. Klausen, L. Hedrich, Formal verification for nonlinear analog systems: approaches to model and equivalence checking, in *Advanced Formal Verification*, ed. by R. Drechsler (Kluwer Academic Publishers, Boston, 2004), pp. 205–245
40. A.T. Davis, An overview of algorithms in Gnucap, in *University/Government/Industry Microelectronics Symposium* (2003), pp. 360–361
41. D. Nickovic, O. Maler, AMT: a property-based monitoring tool for analog systems, in *Formal Modeling and Analysis of Timed Systems*, vol. 4763, Lecture Notes in Computer Science, ed. by J.-F. Raskin, P. Thiagarajan (Springer, Heidelberg, 2007), pp. 304–319
42. R. Alur, C. Courcoubetis, D. Dill, Model-checking in dense real-time. Inf. Comput. **104**, 2–34 (1993)
43. S. Steinhorst, L. Hedrich, Model checking of analog systems using an analog specification language, in *Proceedings of the Design, Automation and Test in Europe DATE '08* (2008), pp. 324–329

44. J.L. Bentley, Multidimensional binary search trees used for associative searching. Commun. ACM **18**(9), 509–517 (1975)
45. B. Chazelle, An optimal convex hull algorithm in any fixed dimension. Discret. Comput. Geom. **10**, 377–409 (1993)
46. M. Althoff, S. Yaldiz, A. Rajhans, X. Li, B. Krogh, L. Pileggi, Formal verification of phase-locked loops using reachability analysis and continuization, in *2011 IEEE/ACM International Conference on Computer-Aided Design (ICCAD)* (IEEE, 2011), pp. 659–666

Author Biographies

Andreas Fürtig received the Diploma degree in computer science from the Goethe-University of Frankfurt in 2013. He is currently pursuing as a Ph.D. student at the institute of computer science within the design methodology group supervised by Prof. Dr. L. Hedrich. His current research interests include formal verification methods, simulation and coverage metrics of analog circuits, as well as behavioral modeling. He is also the teamleader of the RoboCup SPL Team Bembelbots.

Lars Hedrich received the Diploma degree in electrical engineering in 1992 and the Ph.D. from the University of Hannover in 1997 and became a Juniorprofessor at the same University in 2002. Since 2004 he is full professor at the Institute of Computer-Science, University of Frankfurt, and head of the design methodology group at the same institute. His research interests include several areas of analog design automation: symbolic analysis of linear and nonlinear circuits, behavioral modeling, reliability analysis and design, circuit synthesis, and formal verification.

Chapter 2
Verification of Incomplete Designs

Bernd Becker, Christoph Scholl and Ralf Wimmer

2.1 Introduction

With a hierarchical, component based design style unknown values increasingly emerge in different phases of the design and production process, and have to be handled by corresponding electronic design automation (EDA) tools. In the following we restrict our attention to the verification process and describe current state-of-the-art approaches that enable the handling of such unknown values using formal techniques: We provide a sequence of filters, trading off quality, and computation times of the analysis.

Unknown values in circuit verification occur, for instance, when a circuit is only partially available. Partially available means that for some of the circuit's components only their interface is known, i.e., the signals connected to the inputs and outputs of the components, but neither their internal structure nor the computed function. These missing parts are called *black boxes*. The actual values at their outputs are unknown. Verification has to take this into account.

There are different reasons for considering such incomplete circuits: Errors in a circuit design should be detected as early as possible; the later errors are corrected the higher are the incurred costs. Therefore it is desirable to apply verification techniques already in an early stage of the design process when not all parts of a circuit have been implemented yet.

B. Becker (✉) · C. Scholl · R. Wimmer
Institute of Computer Science, Albert-Ludwigs-Universität Freiburg,
Freiburg im Breisgau, Germany
e-mail: becker@informatik.uni-freiburg.de

C. Scholl
e-mail: scholl@informatik.uni-freiburg.de

R. Wimmer
e-mail: wimmer@informatik.uni-freiburg.de

R. Drechsler (ed.), *Formal System Verification*, DOI 10.1007/978-3-319-57685-5_2

A further reason for considering incomplete circuits is that some modules like multipliers are notoriously hard to verify: If the property to be checked is expected to be independent of such a module, the module can be removed from the circuit, and instead it is checked whether the property under consideration holds for all possible replacements of the missing part. If this is the case, then the property also holds for the complete circuit. Otherwise either the remaining circuit is faulty or the removed module and the property interact in some unexpected way.

Considering incomplete circuits can also be beneficial for error diagnosis during debugging. Assume that an error is contained in one of the circuit's modules, but it is not known in which one. If, after removing one module, verification yields that there is an implementation of the removed part such that the considered property holds, then it is likely that the error is contained in the removed module.

If error diagnosis and error rectification are performed late in the design cycle when already a lot of efforts have been made to perform logic synthesis or even place and route steps for the complete design, then the question will be whether the design can be rectified by changing *locally* confined black boxes only, without introducing new connections to global signals leading to enormous costs for re-synthesis. A similar situation occurs in case of Engineering Change Order (ECO, small changes of specification late in the design cycle) where only locally confined parts (black boxes) should be replaced in order to satisfy the changed specification without sacrificing too much of the design efforts. In this case, it is particularly important to preserve the interface of the black boxes.

The synthesis of digital controllers [3, 4] that ensure certain properties of the system at hand can also be considered as a black-box verification problem: The controller to be synthesized is the black box, and one asks whether there is an implementation such that the given property holds.

Depending on the application there are two different problem classes that are of interest: On the one hand, *realizability* asks whether there is an implementation of the black boxes such that the complete design fulfills a certain property. On the other hand, *validity* asks whether the property holds no matter how the missing parts are implemented. Since validity and realizability are dual properties—a property φ is valid iff $\neg\varphi$ is not realizable—in the following we concentrate on realizability problems.

Our methods are based on checking the satisfiability of (quantified) Boolean formulas. In the meantime software tools checking the satisfiability of propositional Boolean formulas are well developed and have been proven to be very successful in many industrial applications [2]. In contrast to propositional formulae used for SAT, quantified Boolean formulae, where variables are existentially or universally quantified, allow a (more) succinct representation in many cases. The recent advances in the performance of QBF solvers, for example, conflict driven learning, resolution and expansion based algorithms [1], or preprocessing [16] enable more exact reasoning about realizability in presence of black boxes even for larger circuits. To be even more exact (or complete) in the general case dependencies between the quantified variables have to be taken into account. This leads to an extension of QBF, called *Dependency quantified Boolean formulas* (DQBF). DQBF allows existential variables to depend on arbitrary sets of universal variables. DQBF solving currently is under development, according to the literature there exist currently two DQBF solvers: IDQ [12] and our tool HQS [15].

We now turn to the verification problems considered in this contribution and at first provide an overview of the problems and algorithms to be presented in more detail in the following sections.

The problem whether an incomplete combinational circuit can be completed such that it becomes equivalent to a given specification (*partial equivalence checking*, PEC) was first considered in [27] where several *approximate* and *exact* methods to solve the PEC problem have been presented. If an approximate algorithm reports that there is no implementation for the black boxes such that the specification holds, the desired specification is indeed not realizable. However, if such an algorithm is not able to prove non-realizability, this can be due to the approximate nature of the method, and the desired functionality may nevertheless be *not* realizable. The algorithms in [27] are based on solving SAT or QBF formulations of PEC. The SAT formulations are efficient to solve also due to a "near at hand" encoding, but also rather inaccurate due to a coarse approximation. Their accuracy is improved in several steps, leading to a QBF formulation that can solve PEC for a single black box exactly. In [27] additionally an exact characterization of realizability of PEC for multiple black boxes has been proposed (based on the decomposability of a certain Boolean relation). However, no feasible algorithmic method for solving the problem has been given.

Nevertheless, [27] was the first paper to consider an exact solution of the PEC problem taking into account that the *interfaces* of the black boxes in the incomplete circuit have to be preserved. Apart from the approach in [18, 19], the diagnosis and rectification problem respecting local interfaces has not been addressed in the literature so far. In [28, 29], e.g., rectifications are computed, but they are allowed to depend on arbitrary signals in the circuit. (Moreover, in contrast to [28, 29] uses a SAT formulation to compute rectifications for a given set of counterexamples only, without considering correctness for *all* possible inputs.) The approach of [18, 19] solves the PEC problem exactly, but it is restricted to problem instances of moderate sizes, since the black boxes are replaced by function tables using an exponential number of Boolean variables. A more efficient complete approach, is based on solving an extension of quantified Boolean formulas, so-called *dependency quantified Boolean formulas (DQBFs)* and was first presented in [14].

We have extended the application of realizability checking to sequential circuits which are specified by a set of properties (safety properties or more general properties formulated in Computation Tree Logic (CTL [9]). Here the question is whether an incomplete sequential design may be extended by black box implementations such that a set of given properties is satisfied. Also the problem of deciding validity is considered. We developed various approaches for solving the realizability problem either in an approximate or an exact manner. In the following, we discuss some representative approaches:

These approaches rely on a series of approximate methods with varying precision and costs for deciding the realizability of properties using symbolic methods. As in the combinational case, the approximations are based on different methods to model the effect of the unknowns at the black box outputs to the overall circuit.

In [22] approximation methods are applied in the context of realizability checking of safety properties based on bounded model checking techniques (BMC—here a sequential circuit is "unrolled" for a number of time frames). This approximate approach leads to SAT or QBF problems. Here, the precision of modeling is not given by the user, but it is adapted automatically based on the difficulty of the problem. The approach is guided by proofs that non-realizability cannot be shown using the weaker methods, independently from the number of BMC unrollings, i.e., independently from the length of a counterexample which does not depend on the implementation of the black boxes. [23] extends [22] and provides proofs based on inductive arguments, showing that non-realizability can not be proven even by our most exact QBF based methods (also independently from the number of BMC unrollings). The approach of [23] provides an exact decision procedure for realizability in the case that the design contains exactly one black box which is allowed to read all input signals (which means that it has "complete information").

The usage of DQBF formulations for verifying incomplete designs was first proposed in [13, 14]. There we have shown that DQBF allows to check realizability precisely for combinational circuits with an arbitrary number of black boxes. This was generalized in to sequential circuits with combinational or bounded-memory black boxes.

In [24], we provided a series of approximate methods for deciding the realizability of CTL properties using symbolic methods. Moreover, [24] presents an exact method for deciding realizability for incomplete circuits with several black boxes under the assumption that the black boxes may contain only a bounded amount of memory. This exact method is based on introducing an exponential number of new variables and is therefore only suitable for small problem instances.

The remaining part of our contribution is structured as follows: In the following section, we introduce the necessary foundations that are required for verifiying incomplete circuits. In Sect. 2.3, we consider the case of incomplete combinational circuits, which do not contain any memory elements. The following Sect. 2.4 considers incomplete sequential circuits. It encompasses two parts: in Sect. 2.4.1 we consider techniques that use bounded model checking to refute realizability. Section 2.4.2 describes OBDD-based model checking algorithms for CTL properties. Finally, in Sect. 2.5 we conclude this paper.

2.2 Preliminaries

In this section, we provide a short overview of the underlying calculus used to solve the verification problems considered in the following. The interested reader is referred to [2] for more details.

Our algorithms are based on checking the satisfiability of (quantified) Boolean formulas. Even though deciding the satisfiability of a quantifier-free Boolean formula

(SAT) is an NP-complete problem [10], SAT solvers, i.e., software tools checking the satisfiability of quantifier-free Boolean formulas have been proven to be very successful in many industrial applications. The formula is typically provided in conjunctive normal form (CNF). A formula in CNF is a conjunction of clauses, and a clause is a disjunction of literals, where a literal is either a Boolean variable or its negation. For instance a clause is given by $(x \vee \neg y)$ with the Boolean variables x and y.

It will turn out that, depending on the problem considered, SAT may not allow an adequate concise description, and extensions like quantified Boolean formulas are more suitable:

Definition 2.1 (*Quantified Boolean formulas: Syntax*) Let V be a set of Boolean variables. A *quantified Boolean formula (QBF)* Ψ (in prenex form) has the form

$$\forall X_1 \exists Y_1 \forall X_2 \ldots \exists Y_k : \varphi,$$

where $X_1, Y_1, X_2, \ldots, Y_k$ are pairwise disjoint subsets of V such that $\bigcup_{i=1}^{k} X_i \cup Y_i = V$, $X_i \neq \varnothing$ for $i = 2, \ldots, k$, and $Y_j \neq \varnothing$ for $j = 1, \ldots, k-1$. φ is a Boolean formula over V, called the *matrix* of Ψ. $\forall X_1 \exists Y_1 \forall X_2 \ldots \exists Y_k$ is called Ψ's *quantifier prefix*.

Typically the matrix φ of a QBF is required to be in conjunctive normal form. A QBF has a linearly ordered quantifier prefix: each existential variable depends on all universal variables in whose scope it is. If we consider a QBF as a two-player game, then one player assigns the existential variables and the other player the universal ones. The game proceeds according to the quantifier prefix from left to right. The goal of the existential player is to satisfy the matrix, the universal player's goal to falsify it. Thereby, each player may base the assignment of the current variable on all assignments the opponent has made so far. The QBF is satisfied iff the existential player has a winning strategy which satisfies the matrix for all possible assignments of the universal variables.

Example 2.1 As an example consider the QBF $\forall x_1 \exists y_1 \forall x_2 \exists y_2 : \varphi$. It is satisfied if and only if: for *every* assignment of x_1, there *exists* an assignment for y_1 such that for *every* assignment of x_2 there *exists* an assignment for y_2, such that the matrix is satisfied. Obviously, here y_2 depends on x_2 and on x_1.

The complexity of QBF satisfiability is determined by the number of quantifier alternations between existential and universal quantifiers and vice versa in the prenex form. The general problem of QBF satisfiability is a PSPACE-complete problem [21].

In a next step, the dependencies between the quantified variables can be generalized: As already mentioned, in QBF each existential variable depends on all universal variables to the left. *Dependency quantified Boolean formulas* (DQBF) relax this restriction and allow existential variables to depend on arbitrary sets of universal variables.

Definition 2.2 (*Dependency quantified Boolean formulas: Syntax*) Let $V = \{x_1, \ldots, x_n, y_1, \ldots, y_m\}$ be a set of Boolean variables and φ a quantifier-free Boolean formula over V. A *dependency quantified Boolean formula (DQBF)* ψ over V has the form

$$\psi = \forall x_1 \forall x_2 \ldots \forall x_n \exists y_y(D_{y_1}) \exists y_2(D_{y_2}) \ldots \exists y_m(D_{y_m}) : \varphi,$$

where $D_{y_i} \subseteq \{x_1, \ldots, x_n\}$ is the dependency set of y_i.

The semantics of such a formula is typically defined in terms of Skolem functions. For a set V of Boolean variables, let $\mathscr{A}(V) = \{\nu \,|\, \nu : V \to \{0, 1\}\}$ denote the set of all variable assignments for V.

Definition 2.3 (*Dependency quantified Boolean formulas: Semantics*) Let ψ be a DQBF as defined above. It is *satisfied* if there are functions $s_{y_i} : \mathscr{A}(D_{y_i}) \to \mathbb{B}$ such that replacing all occurrences of y_i in φ by (a Boolean expression for) s_{y_i} for all $i = 1, \ldots, m$ turns φ into a tautology.

Example 2.2 Consider the following DQBF (with an appropriate matrix φ):

$$\forall x_1 \forall x_2 \exists y_1(x_1) \exists y_2(x_2) : \varphi \,.$$

Here y_1 depends only on x_1 and y_2 only on x_2. There is no QBF prefix which expresses the same dependencies.

We will come back to this example in the following section where we will see that DQBF is an adequate formalism to model verification problems for incomplete combinational circuits.

In general, DQBF has a strictly higher expressiveness than QBF. This is also reflected in the complexity of the decision problem: deciding whether a DQBF is satisfiable is NEXPTIME-complete [25].

First DQBF solvers are currently under development: IDQ was the first available DQBF solver. It applies instantiation-based solving [12]. Our tool HQS [15] is currently the only other available DQBF solver. It uses quantifier elimination to turn the DQBF at hand into an equisatisfiable QBF, which can be solved by an arbitrary QBF solver. To reduce the computation times, HQS applies sophisticated preprocessing techniques. It is not only able to decide the satisfiability of DQBFs, but also to compute—in case the formula is satisfiable—Skolem functions for the existential variables [31].

2.3 Incomplete Combinational Circuits

In this chapter, we are going to investigate the verification of incomplete combinational circuits, i.e., circuits without memory elements containing unknown parts. After defining the partial equivalence checking problem, we investigate symbolic

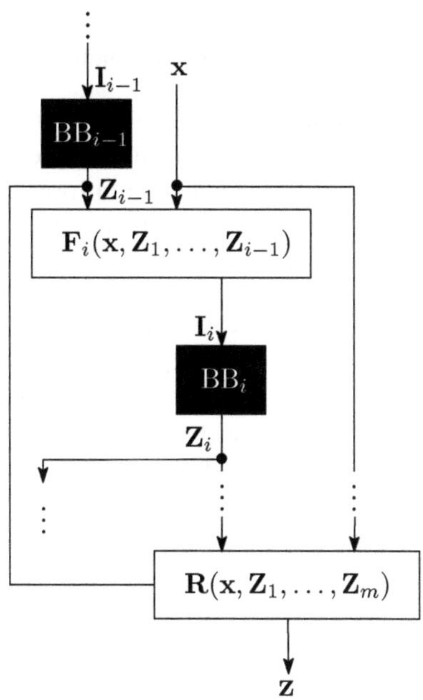

Fig. 2.1 Notation for incomplete combinational circuits

simulation with 01X-logic. We provide a sound formulation that enables an efficient, but incomplete solution method: If an answer is obtained, it is correct, but there are cases where the procedure cannot give a conclusive answer. We improve on this by using quantified Boolean formulas (QBF). For combinational circuits with only a single black box, the QBF formulation entails a complete decision procedure; for more than one black box, it is still an approximation. In this case, the extension of QBF to dependency quantified Boolean formulas (DQBF) has to be used.

2.3.1 *The Partial Equivalence Checking Problem (PEC)*

We first define incomplete circuits and the partial equivalence checking problem.

Figure 2.1 shows a part of an incomplete circuit. We assume that $\mathbf{x} = (x_1, \ldots, x_n)$ are the primary inputs of the circuit,[1] $BB_1, \ldots, BB_m$ are its unknown parts (so-called "black boxes"). In order to guarantee that the circuit is combinational for all possible combinational black box implementations, we assume that the black boxes are given in topological order, i.e., BB_i does not depend on BB_j for $1 \le i < j \le m$. Furthermore, we assume that the interfaces of the black boxes are known: $\mathbf{I}_i$ is the vector of input signals of BB_i and $\mathbf{Z}_i$ the vector of its output signals for $i = 1, \ldots, m$.

[1] In the following, we denote individual signals or variables in italic font (like x, y, z) and vectors in upright bold font (like $\mathbf{x}$, $\mathbf{y}$, $\mathbf{z}$).

The known parts of the circuit are given by the vectors $\mathbf{F}_i(\mathbf{x}, \mathbf{Z}_1, \ldots, \mathbf{Z}_{i-1})$, which define $\mathbf{I}_i$, and the output functions $\mathbf{z} = \mathbf{R}(\mathbf{x}, \mathbf{Z}_1, \ldots, \mathbf{Z}_m)$ of the circuit.

Definition 2.4 (*Partial Equivalence Checking Problem*) The *Partial Equivalence Checking (PEC)* problem is defined as follows: Given an incomplete combinational circuit and a specification in form of a complete circuit, are there implementations of the black boxes such that both circuits become equivalent? In this case, the specification is said to be *realizable*.

Equivalent means that both circuits produce, for the same input pattern, the same output values. By a (negated) miter construction, we can combine the two circuits into a single one: corresponding inputs are driven by the same input signal, corresponding outputs are connected via equivalence gates (xnor-gates), and their outputs in turn via an and-gate. The single output of this miter circuit is 1 for an input pattern if both circuits compute the same value. The specification is unrealizable iff for all possible black box implementations there is an input pattern such that the output of the negated miter becomes 0.

In the following, we assume that the PEC is given as the negated miter circuit with the goal to make the output constantly 1 by implementing the black boxes appropriately.

2.3.2 SAT-based Approximations

One possibility to model the unknown behavior of the black boxes is to use a three-valued logic (also called 01X-logic), which adds a third value X to the classical Boolean logic. X stands for an unknown value. The Boolean operations can be generalized easily to the three-valued logic, e.g., by $0 \wedge \mathrm{X} = 0$, $1 \wedge X = \mathrm{X}$, etc. The complete truth tables for $\wedge$, $\vee$, and $\neg$ for the three-valued logic are shown in Table 2.1. Given a Boolean assignment for the primary inputs, we can assign X to the outputs of the black boxes and do a simulation of the circuit using the truth tables in Table 2.1. If the value of a circuit output is 0 or 1, this value is independent of the implementation of the black boxes. Obtaining output X means that, for the given input pattern, the output might depend on the black box implementation.

01X-logic is known to over-estimate the number of X-values in a circuit. Consider, for example, a circuit with one and- and one not-gate that computes $a \wedge (\neg a)$. The output of this circuit is 0 for all possible values of a, but in 01X-logic, if $a = \mathrm{X}$, the output is determined to be X.

Still, we can use 01X-logic to determine unrealizability of a PEC problem. To do so, we use symbolic simulation using the three-valued logic: Take the formula $z = R(\mathbf{x}, \mathbf{Z}_1, \ldots, \mathbf{Z}_m)$, which defines the output of the circuit, and check if, given an assignment of X to all black box outputs, there exists an assignment of $\mathbf{x}$ (in $\{0, 1\}$) such that z has the value 0 (or equivalently $\neg z$ value 1):

Table 2.1 Boolean operations on 01X-logic

(a) Conjunction

$\wedge$	0	1	X
0	0	0	0
1	0	1	X
X	0	X	X

(b) Disjunction

$\vee$	0	1	X
0	0	1	X
1	1	1	1
X	X	1	X

(c) Negation

$\neg$	0	1	X
	1	0	X

$$\big(z \equiv R(\mathbf{x}, \mathbf{Z}_1, \ldots, \mathbf{Z}_m)\big) \wedge \big(x_1 \neq \mathrm{X}\big) \wedge \cdots \wedge \big(x_n \neq \mathrm{X}\big)$$
$$\wedge \big(\mathbf{Z}_1 = \mathbf{X}\big) \wedge \cdots \wedge \big(\mathbf{Z}_m = \mathbf{X}\big) \wedge \big(z = 0\big).$$

This is a satisfiability (SAT) problem over 01X-logic. To make standard SAT solvers applicable, we have to re-encode the formula over standard Boolean logic. Jain et al. [17] propose to use $\lceil \log_2 3 \rceil = 2$ Boolean variables to encode the three different values 0, 1, and X: 0 is encoded as (1, 0), 1 as (0, 1), and X as (0, 0). For three-valued variables a, b with encodings (a^0, a^1) and (b^0, b^1), resp., we obtain: $a \vee b = (a^0 \wedge b^0, a^1 \vee b^1)$, $a \wedge b = (a^0 \vee b^0, a^1 \wedge b^1)$, and $\neg a = (a^1, a^0)$. Of course, this encoding has to be applied only to those signals which are in the cone of influence of the black box outputs. As a Boolean signal a is equivalent to the encoding $(\neg a, a)$, we can convert standard Boolean signals into encoded three-valued ones where necessary.

The result of encoding the three-valued miter circuit is a Boolean circuit with two outputs (z^0, z^1) with $z^0 \equiv R^0(\mathbf{x}^0, \mathbf{x}^1, \mathbf{Z}_1^0, \mathbf{Z}_1^1, \ldots, \mathbf{Z}_m^0, \mathbf{Z}_m^1)$ and $z^1 \equiv R^1(\mathbf{x}^0, \mathbf{x}^1, \mathbf{Z}_1^0, \mathbf{Z}_1^1, \ldots, \mathbf{Z}_m^0, \mathbf{Z}_m^1)$. By adding appropriate clauses to the SAT formula, we have to enforce that the following restrictions are fulfilled: (1) $z^0 \wedge \neg z^1$, i.e., $(z^0, z^1) = (1, 0)$, (2) $\neg Z_{i,j}^0 \wedge \neg Z_{i,j}^1$ for $i = 1, \ldots, m$ and $Z_{i,j} \in \mathbf{Z}_i$, i.e., all black box outputs are assigned to X, and (3) $x_i^0 \oplus x_i^1 = 1$ for $i = 1, \ldots, n$, i.e., $(x_i^0, x_i^1) \in \{(0, 1), (1, 0)\}$.

If the resulting Boolean formula is satisfiable, there is an input pattern for the circuit (which corresponds to the assignment of $(x_1^1, \ldots, x_n^1)$) that leads to an output of $0 = (1, 0)$, independent of the actual implementation of the black boxes. Hence, the PEC is unrealizable. Contrary, if the formula is unsatisfiable, we cannot conclude realizability: We cannot distinguish whether the design is realizable or the applied method too weak, caused by one of the following limitations:

- The 01X-logic is not precise in case of reconvergences of the circuit.
- Using 01X-logic, we search for counterexample patterns (refuting realizability) that are independent of the black boxes. However, in general, we might need a different pattern for each implementation of the black boxes.
- We ignore the input cones of the black boxes.

This is improved step by step in the following section by using quantified Boolean formulas.

2.3.3 QBF-based Methods

The problem that 01X-logic is inaccurate in case of reconvergences can be solved by using quantified Boolean formulas. The idea is to check if for all assignments of the primary inputs there are values of the black box outputs such that the formula $R(\mathbf{x}, \mathbf{Z}_1, \ldots, \mathbf{Z}_m)$ evaluates to 1. This yields the following QBF:

$$\forall \mathbf{x} \exists \mathbf{Z}_1 \ldots \exists \mathbf{Z}_m : \big(z \equiv R(\mathbf{x}, \mathbf{Z}_1, \ldots, \mathbf{Z}_m)\big) \wedge \big(z = 1\big). \tag{2.1}$$

Unsatisfiability of this formula implies unrealizability of the PEC problem. This formulation is more precise than the 01X-based formulation, but typically more expensive to solve—while SAT is NP-complete [10], QBF is PSPACE-complete [21]. This also affects the solver performance from a practical point of view: SAT problems that can be solved nowadays may be roughly two orders of magnitude larger than solvable QBFs.

In spite of the higher precision, (2.1) still does not constitute a complete decision method for PEC: so far, we have neglected the input cones of the black boxes.

If the black boxes are not directly connected to the primary inputs but to internal signals we have to take into account that not all possible combinations of values may arrive at the inputs of the black boxes, i.e., that the black boxes have only partial information about the primary inputs.

Since we use universal quantification for the black box inputs, we only have to ensure that our formula is satisfied, if the value of the black box inputs $\mathbf{I}_i$ does not deviate from the values obtained as a function $\mathbf{I}_i = \mathbf{F}_i(\mathbf{x}, \mathbf{Z}_1, \ldots, \mathbf{Z}_{i-1})$. The following matrix fulfills this requirement:

$$\varphi = \big(\mathbf{I}_1 \not\equiv \mathbf{F}_1(\mathbf{x})\big) \vee \cdots \vee \big(\mathbf{I}_m \not\equiv \mathbf{F}_m(\mathbf{x}, \mathbf{Z}_1, \ldots, \mathbf{Z}_{m-1})\big) \vee R(\mathbf{x}, \mathbf{Z}_1, \ldots, \mathbf{Z}_m)\,. \tag{2.2}$$

We use universal quantification for the primary inputs $\mathbf{x}$ and all black box inputs $\mathbf{I}_1, \ldots, \mathbf{I}_m$. The outputs $\mathbf{Z}_1, \ldots, \mathbf{Z}_m$ of the black boxes are existentially quantified. If a black box output is the input of another black box, we introduce a buffer, "computing" the identity function, to avoid conflicts. We have to take care that in the quantifier prefix, all inputs of a black box precede its outputs. Then there are still several prefixes possible, e.g.,

$$\forall \mathbf{x} \forall \mathbf{I}_1 \ldots \forall \mathbf{I}_m \exists \mathbf{Z}_1 \ldots \exists \mathbf{Z}_m : \varphi, \quad (2.3)$$

$$\forall \mathbf{I}_1 \exists \mathbf{Z}_1 \forall \mathbf{I}_2 \backslash \mathbf{I}_1 \exists \mathbf{Z}_2 \ldots \exists \mathbf{Z}_m . \forall \mathbf{x} \backslash (\mathbf{I}_1 \cup \mathbf{I}_2 \cup \cdots \cup \mathbf{I}_m) : \varphi \quad (2.4)$$

The QBF with prefix (2.3) contains only two quantifier blocks. It is therefore typically easier to solve than (2.4). The drawback is that is less precise: In a QBF, an existential variable depends on all universal ones left of it in the prefix. In (2.4), each existential variable in general depends on fewer variables than in (2.3). Additionally, (2.4) is precise in case of a single black box. For more than one black box, (2.4) is not able to express the dependencies of the black box outputs on their corresponding inputs exactly.

2.3.4 *DQBF-based Methods*

In the previous two sections, we made the decision procedure for PEC more and more precise. One last step is missing in order to obtain a complete and sound decision procedure for PEC.

If a combinational design contains more than one black box and if the black boxes do not have the same interfaces, it is typically not possible to express the dependencies of the black boxes' outputs on their inputs in QBF exactly. This is illustrated in the following example.

Example 2.3 Consider an incomplete design with two black boxes BB_1 and BB_2. The output Z_1 of black box BB_1 depends only on x_1, the output Z_2 of BB_2 only on x_2.[2] Then three different QBF formulation are possible (with an appropriate matrix φ): (1) $\forall x_1 \exists Z_1 \forall x_2 \exists Z_2 : \varphi$, (2) $\forall x_2 \exists Z_2 \forall x_1 \exists Z_1 : \varphi$, and (3) $\forall x_1 \forall x_2 \exists Z_1 \exists Z_2 : \varphi$, In formula (1), Z_2 depends not only on x_2, but also on x_1, violating the interface specification of BB_2. The same holds in (2) for BB_1, which would be allowed to read x_2. Finally, in (3) both black boxes read both signals.

To be able to express the dependencies of the black box outputs exactly, we have to resort to *dependency quantified Boolean formulas (DQBFs)*, as already introduced in the previous section. While a QBF has a linearly ordered quantifier prefix—each existential variable depends on all universal variables in whose scope it is—in a DQBF the existential variables are explicitly annotated with the universal variables they depend upon.

[2] In this example we assume that the inputs of the black boxes are directly connected to the primary inputs. Thus we do not need to introduce separate vectors $\mathbf{I}_1$ and $\mathbf{I}_2$ for the black box inputs.

Example 2.4 Consider again the incomplete design from Example 2.3. It leads to the following DQBF (with an appropriate matrix φ):

$$\forall x_1 \forall x_2 \exists Z_1(x_1) \exists Z_2(x_2) : \varphi .$$

Here Z_1 depends on x_1 and Z_2 on x_2, which coincides with the interface specification of the incomplete design.

In general, we can use the following DQBF formulation for PEC:

$$\psi = \forall \mathbf{x} \forall \mathbf{I}_1 \ldots \forall \mathbf{I}_m \exists \mathbf{Z}_1(\mathbf{I}_1) \ldots \exists \mathbf{Z}_m(\mathbf{I}_m) : \varphi, \tag{2.5}$$

where the matrix φ is exactly the same as in Eq. (2.2) defined for QBF. In contrast to QBF, DQBF allows us to model arbitrary black box interfaces precisely.

It is not hard to show that there is not only a linear transformation from PEC to DQBF, but also vice versa [13, 14]. The existence of these transformations implies on the one hand that PEC and DQBF have the same complexity. They are both NEXPTIME-complete. On the other hand, we can solve PEC exactly by solving an appropriate DQBF formulation.

2.4 Incomplete Sequential Circuits

We investigate how and to which extent the techniques presented in the previous section for incomplete combinational circuits can be generalized to incomplete circuits, which contain memory elements, i.e., incomplete sequential circuits.

In contrast to combinational circuits, we consider the validity and realizability of a property regarding an incomplete sequential circuit and then present approaches for the analysis of those circuits.

Analogously to the combinational case an *incomplete sequential circuit* is a sequential circuit containing black boxes. See Fig. 2.2 for an illustration. The circuit contains m black boxes $\mathrm{BB}_1, \ldots, \mathrm{BB}_m$, shown as black rectangles. Their input signals are denoted by $\mathbf{I}_1, \ldots, \mathbf{I}_m$ and their output signals by $\mathbf{Z}_1, \ldots, \mathbf{Z}_m$. The primary inputs of the circuit are $\mathbf{x} = (x_0, \ldots, x_n)$, the current state is given by the signals $\mathbf{s} = (s_0, \ldots, s_r)$. The input cones of the black boxes compute the functions $\mathbf{I}_i = \mathbf{F}_i(\mathbf{x}, \mathbf{s}, \mathbf{Z}_1, \ldots, \mathbf{Z}_{i-1})$. We thereby assume that there are no cyclic dependencies between the black boxes and that they are topologically ordered, i.e., BB_i only depends on the values computed by $\mathrm{BB}_1, \ldots, \mathrm{BB}_{i-1}$. To simplify notation, we assume w. l. o. g. that no black box output is directly connected to a black box input, i.e., $\mathbf{Z}_i$ is disjoint from $\mathbf{I}_j$ for all i, j. If this is not the case, we insert a buffer between the corresponding black boxes, which does not modify the functionality of the circuit. Finally, the output $\mathbf{y}$ and the next state $\mathbf{s}'$ of the circuit are given by the Boolean functions $(\mathbf{y}, \mathbf{s}') = \mathbf{R}(\mathbf{x}, \mathbf{s}, \mathbf{Z}_1, \ldots, \mathbf{Z}_m)$.

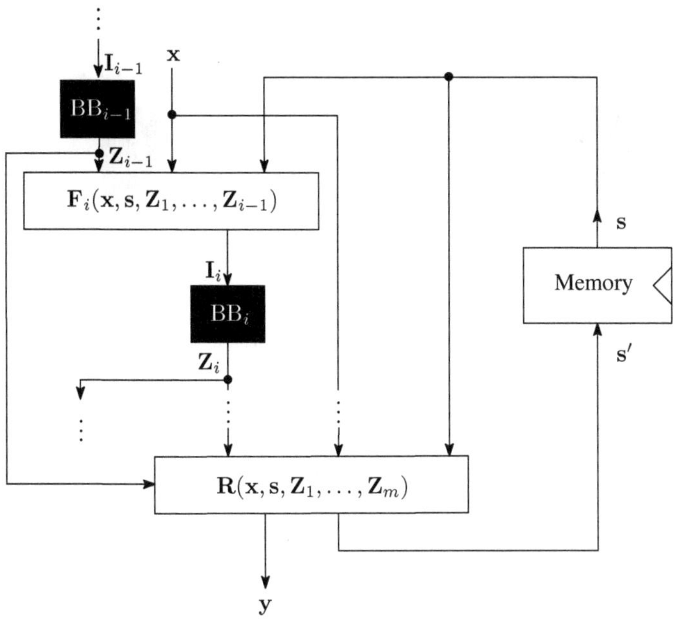

Fig. 2.2 Notations for an incomplete sequential circuit

The complexity of deciding realizability depends on the allowed behavior of the black boxes: If we assume that the contents of the black boxes are combinational circuits, the problem of deciding realizability is NEXPTIME-complete. If we allow the black boxes to contain an arbitrary amount of memory, the interesting decision problems (see below) become undecidable [26]. The case of black boxes with a bounded amount of memory (i.e., we know an upper bound on the number of bits the black boxes can store internally) can be reduced to the case of combinational black boxes by adding the memory of the black boxes to the surrounding circuit such that these memory cells are read and written only by the corresponding black box.

For a given property P two questions regarding a partial circuit are of interest: On the one hand, *realizability* asks whether there is an implementation of the black boxes such that the complete circuit satisfies P. On the other hand, *validity* asks whether P is satisfied for all possible implementations. Since validity of P is given iff $\neg P$ is not realizable, we restrict ourselves in the following to realizability problems.

In particular, in the following subsection we present approaches based on bounded model checking techniques (BMC) [22, 23]. As properties we consider invariant properties: Given a Boolean formula $P(\mathbf{x}, \mathbf{s}, \mathbf{y})$, which describes the states of the circuit that satisfy the invariant, are there implementations of the black boxes such that $P(\mathbf{x}, \mathbf{s}, \mathbf{y})$ is satisfied in each step of the circuit? For more general classes of properties like arbitrary CTL properties, we refer the reader to Subsect. 2.4.2.

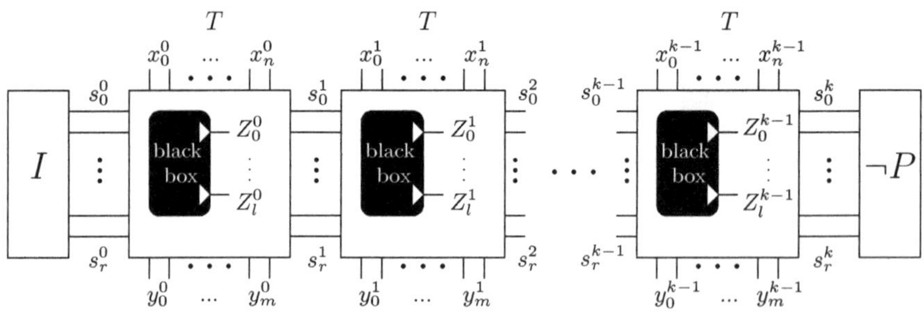

Fig. 2.3 Encoding of the BMC problem for incomplete designs [23]

2.4.1 *BMC for Incomplete Designs*

BMC for incomplete designs aims to refute the realizability of a property, that is, it tells the designer, no matter how the unknown parts of the system will be implemented, the property will always fail. To put it in other words, the error is already in the implemented system. If this is the case, then we call the property P *unrealizable*. Here we restrict the properties to *invariants*. In a first formulation, we make use of QBF modeling where the variables representing the black box outputs are universally quantified. We even allow black box replacements to produce different output values for the same input values at different time steps which simplify the formula but is a source of inexactness in case that only combinational circuits are allowed as black box implementations and also in case of black boxes with a bounded amount of memory[3]—and thus the method might miss to prove the unrealizability of some properties. Moreover, we first confine ourselves to single black boxes. If several black boxes with dedicated interfaces are combined into a single black box, then this is another source of inexactness. To encode the BMC problem of incomplete designs, we are naming the variables as shown in Fig. 2.3. We use an upper index to specify the time instance of a variable. s_i^j denotes the i-th state bit in the j-th unfolding (let $\mathbf{s}^j = s_0^j, \ldots, s_r^j$). The same holds for the primary inputs $\mathbf{x}^j = x_0^j, \ldots, x_n^j$, the primary outputs $\mathbf{y}^j = y_0^j, \ldots, y_m^j$, and the black box outputs $\mathbf{Z}^j = Z_0^j, \ldots, Z_l^j$. The next state variables $\mathbf{s}^{j+1}$ depend on the current state, the primary inputs and the black box outputs. The whole circuit is transformed according to [30] using additional auxiliary variables $\mathbf{H}^j$ for each unfolding depth j. The predicate describing the initial states is given by $I(\mathbf{s}^0)$. Since we assume a single initial state in this paper, the initial state $I(\mathbf{s}^0)$ is encoded by unit clauses, setting the respective state bits to their initial value. The transition relation of time frame i is given by $T(\mathbf{s}^{i-1}, \mathbf{x}^{i-1}, \mathbf{Z}^{i-1}, \mathbf{s}^i)$. The invariant $P(\mathbf{s}^k)$ is a Boolean expression over the state variables[4] of the k-th unfolding.

[3]Black boxes with a bounded amount of memory are reduced to the case of combinational black boxes by adding the memory of the black boxes to the surrounding circuit as mentioned above.

[4]In general the property can also check the primary inputs and outputs, but for sake of convenience, we omit details here.

Using this information, the quantifier prefix (and the matrix) for the unrealizability problem results in the QBF formula (2.6). For the sake of simplicity we include the variables representing the primary outputs of unfolding depth j into $\mathbf{H}^j$.

$$
\begin{aligned}
BMC(k) \; := \; & \exists\, \mathbf{s}^0\, \mathbf{x}^0 \quad \forall\, \mathbf{Z}^0 \quad \exists\, \mathbf{H}^0 \\
& \exists\, \mathbf{s}^1\, \mathbf{x}^1 \quad \forall\, \mathbf{Z}^1 \quad \exists\, \mathbf{H}^1 \\
& \qquad \vdots \\
& \exists\, \mathbf{s}^{k-1}\, \mathbf{x}^{k-1}\, \forall\, \mathbf{Z}^{k-1}\, \exists\, \mathbf{H}^{k-1} \\
& \exists\, \mathbf{s}^k
\end{aligned}
$$

$$
I(\mathbf{s}^0) \wedge \bigwedge_{i=1}^{k} T(\mathbf{s}^{i-1}, \mathbf{x}^{i-1}, \mathbf{Z}^{i-1}, \mathbf{s}^i) \wedge \neg P(\mathbf{s}^k) \tag{2.6}
$$

The semantics following from the prefix corresponds to the following question:

> Does there exist a state $\mathbf{s}^0 = s_0^0, \ldots, s_r^0$ and an input vector $\mathbf{x}^0 = x_0^0, \ldots, x_n^0$ at depth 0 such that for all possible values of the black box outputs $\mathbf{Z}^0 = Z_0^0, \ldots, Z_m^0$ there exists an assignment to all auxiliary variables $\mathbf{H}^0$ (resulting in a next state $\mathbf{s}^1 = s_0^1, \ldots, s_r^1$) and an input vector $\mathbf{x}^1 = x_0^1, \ldots, x_n^1$ at depth 1, etc. such that the property is violated at time frame k?

The BMC procedure iteratively unfolds the incomplete circuit for $k = 0, \ldots, K$ until a predefined maximal unfolding depth K is reached. If a QBF solver finds $BMC(k)$ satisfiable, the unrealizability of the property P has been proven. In that case, the resulting system can reach a "bad state" after k steps, no matter how the black box is implemented.

We can prove that, whenever $BMC(k)$ is *unsatisfiable*, there is an implementation of the black box (possibly with a limited number of memory elements) which is able to avoid error paths of length k *as long as the black box is allowed to read all primary inputs*. However, if the black box in the design at hand is not directly connected to all primary inputs, (i.e., if the black box does not have "complete information"), such an implementation does not need to exist. Thus, for black boxes having "incomplete information" the property may be unrealizable although BMC with QBF modeling is not able to prove this. In this case, DQBF is necessary to express the actual dependencies of the black boxes on their inputs.

In the following, we give an example illustrating the approach:

Example 2.5 Consider the incomplete circuit shown in Fig. 2.4. The state bits s_0 and s_1 depend on the current state, the primary input x, and the black box outputs Z_0 and Z_1, respectively, and are computed by the transition functions $s_0' = x \vee Z_0$ and $s_1' = (x \wedge Z_1) \vee (s_0 \wedge \neg Z_1)$. Let the invariant property $P = \neg(s_0 \wedge s_1)$ state that s_0 and s_1 must never be 1 at the same time. Let the initial state of the system be defined as $s_0^0 = s_1^0 = 0$. After checking for an initial violation of the property, the BMC procedure unfolds the system once, and tries to find an assignment to x^0 such that for

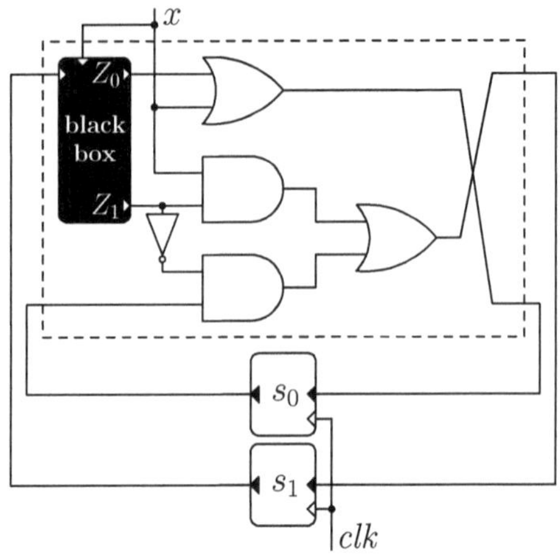

Fig. 2.4 Example of an incomplete design [23]

all possible assignments to Z_0^0 and Z_1^0 the state $(1, 1)$ can be reached. Indeed, $x^0 = 1$ implies $s_0 = 1$ for all assignments to the black box outputs, however, for $Z_1^0 = 0$ $s_1 = 1$ cannot be obtained (neither by setting $x^0 = 0$ nor $x^0 = 1$). Thus, $BMC(1)$ is unsatisfiable and BMC continues by adding a second copy of the transition relation to the problem. If $x^0 = 1$, the current state bit s_0^1 at the second unfolding evaluates to 1 as well. Furthermore, if x^1 is set to 1, the next state bits s_0^2 and s_1^2 evaluate to 1 for all values of Z_0^1 and Z_1^1. Hence, when applying the input pattern $x^0 = x^1 = 1$, a state violating P can be reached after two steps for all actions of the black box and thus, $BMC(2)$ is satisfiable and P is unrealizable.

2.4.1.1 SAT-based Approximations and 01X-Hardness

As already shown in previous sections, QBF can be approximated using 01X-logic. In the previous example, it was not necessary to use QBF encoding for Z_0. When applying $Z_0 = X$, unrealizability still can be proven by applying $x^0 = x^1 = 1$.

Since the problem instances using 01X-modeling are typically easier to solve, we are using the following verification flow:

Given an incomplete design and an invariant, we start the BMC process with a pure 01X-modeling, that is we extend Boolean logic by a third value ‘X’ which then is applied to all black box outputs. Using the two-valued encoding proposed by Jain (cf. Sect. 2.3.2), the BMC unfoldings still yield SAT problems which can be solved by a state-of-the-art SAT solver. However, 01X-modeling may be too coarse to prove unrealizability leading to unsatisfiable BMC instances for every unfolding depth (we call such verification problems *01X-hard*). In [22] we presented a method based on Craig interpolation to classify 01X-hard problems on-the-fly along the BMC process, thus preventing the solver running into unsatisfiable instances forever. Additionally, the computed Craig interpolants provide information about the origin of the 01X-hardness, and a subset of the black box outputs which have to be modeled more precisely using QBF is heuristically determined. Now a QBF-based BMC tool processes the information gathered from the Craig interpolants and uses one universally quantified variable for each black box output which needs a more precise modeling. Using a combined 01X/QBF-modeling (or a pure QBF-modeling) the BMC unfoldings yield QBF formulas. In that way, the precision of modeling is not given by the user, but it is adapted automatically based on the difficulty of the problem.

2.4.1.2 QBF-based Approximations and QBF-hardness

However, even when using the more precise QBF modeling technique to model the unknown behavior of the black box, no conclusive result is guaranteed. At this point, an extension given in [23] is introduced into the workflow. Similar to 01X-hardness for 01X-modeled incomplete designs, a QBF modeled BMC problem can now be classified as *QBF-hard*, if QBF-based BMC would continuously run into unsatisfiable unfoldings.

As already discussed above, under certain conditions (black boxes having "incomplete information") the BMC procedure using a QBF formulation is not able to prove unrealizability even if the property is indeed unrealizable. In this sense, the QBF formulation is a sound but incomplete approximation (just as 01X-modeling which is also an approximation, but is strictly coarser). If unrealizability cannot be proven due to the approximative nature of the method or if the property is really realizable, then the BMC procedure described above would produce unsatisfiable QBF formulas for all unfoldings and would never return a result.

The idea of proving QBF-hardness is as follows: The QBF-based BMC procedure classifies a property as unrealizable iff there exist input sequences of some length k such that independently from the black box actions the property will be violated after k steps. Conversely, the QBF-based BMC procedure is not able to prove unrealizability with an unfolding of length k or smaller, if for each input valuation in each time frame there is an action of the black box such that the property is fulfilled after k steps, and additionally all states on these paths also fulfill the property. Furthermore, if we can prove for this scenario that after at most k steps every state has already been visited before, we can be sure that the QBF-based BMC procedure will *never* produce a satisfiable instance, since for every input pattern it is possible to determine at least one realization of the black box leading to a state which does not violate the property, independently from the length of the unfolding.

This concept is illustrated in Fig. 2.5. Let $\mathbf{s}^0 = \mathbf{\oe}_1^0$ be the initial state which fulfills P. Next, the graph branches for all possible assignments ${}_{,1}^{0}, \ldots, {}_{,m}^{0}$ to the primary inputs $\mathbf{x}^0$. For each of these values ${}_{,i}^{0}$ there exists an action of the black box outputs $\mathbf{Z}^0 = \mathbf{\imath}_i^0$ leading to next states $\mathbf{s}^1 = \mathbf{\oe}_i^1$ which all fulfill P. Once a state is equivalent to a state which was visited before (which is indicated by a dashed backward arrow in Fig. 2.5 stating that $\mathbf{\oe}_1^1 = \mathbf{\oe}_1^0$, $\mathbf{\oe}_1^2 = \mathbf{\oe}_2^1$, $\mathbf{\oe}_m^2 = \mathbf{\oe}_1^0$, respectively), this branch does not need to be further explored. If at some depth all so far explored

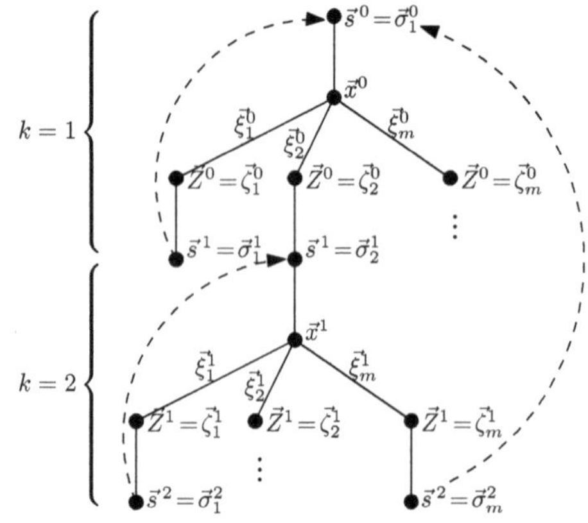

Fig. 2.5 QBF-hardness graph [23]

states point back to already visited states, then the black box outputs are set in a way that the system remains in "good states" forever, i.e., we are in the situation sketched above and we can be sure that the QBF-based BMC procedure will never produce a satisfiable instance, independently from the length of the unfolding. Thus, determining whether a graph fulfilling the aforementioned properties exists answers the question of whether a design is QBF-hard.

In [23] it has been shown that the existence of a QBF-hardness graph can be checked using a series of QBF formulas. Once the QBF-hardness of the design under verification is proven, two options are considered. In case of a combined 01X/QBF-modeling an abstraction refinement procedure will identify more black box outputs for QBF-modeling (in an extreme case yielding a pure QBF-modeled problem) and repeat the QBF-based BMC procedure. If all black box outputs are already QBF-modeled, one has to resort to DQBF modeling.

2.4.1.3 DQBF-based Modeling

If no upper bound on the amount of memory in the black boxes is known, one has to use BMC with DQBF modeling. The dependency sets of the black box outputs are chosen in the following way: If black box BB_i with outputs $\mathbf{Z}_i$ reads the signals $\mathbf{I}_i$, then in the formula $BMC(k)$, the dependency set $D_{\mathbf{Z}_i^j}$ of $\mathbf{Z}_i$ in the j-th unrolling contains all instances of $\mathbf{I}_i$ up to j, i.e., $D_{\mathbf{Z}_i^j} = \{\mathbf{I}_i^k \mid k = 0, \ldots, j\}$. This is exactly the information that has been read by BB_i up to the j-th unrolling. In the worst case, all of this information has been stored in the black boxes' internal memory and the output in the j-th step may depend on it.

Still, this approach is an incomplete decision procedure as the decision problem itself is in general undecidable.

However, if we assume that the black box implementations are combinational circuits, i.e., that the black boxes do not contain any memory elements, realizability can be formulated as a DQBF without the necessity to unroll the circuit. The following formula is satisfiable iff the sequential circuit is realizable with combinational implementations of the black boxes:

$$\forall \mathbf{s}\, \forall \mathbf{s}' \forall \mathbf{x}\, \forall \mathbf{I}_1 \ldots \forall \mathbf{I}_n\, \exists \mathbf{Z}_1(\mathbf{I}_1) \ldots \exists \mathbf{Z}_n(\mathbf{I}_n)\, \exists w(\mathbf{s})\, \exists w'(\mathbf{s}') : \\ \big(I(\mathbf{s}) \Rightarrow w\big) \wedge \big(w \Rightarrow P(\mathbf{s})\big) \wedge \big(\mathbf{s} \equiv \mathbf{s}' \Rightarrow w \equiv w'\big) \wedge \\ \Big(\big(w \wedge \bigwedge_{i=1}^{n} (\mathbf{I}_i \equiv \mathbf{F}_i(\mathbf{x}, \mathbf{s}, \mathbf{Z}_1, \ldots, \mathbf{Z}_{i-1})) \wedge T(\mathbf{s}, \mathbf{x}, \mathbf{Z}_1, \ldots, \mathbf{Z}_n, \mathbf{s}')\big) \Rightarrow w'\Big). \tag{2.7}$$

This formula is based on the notion of a winning set: A subset $W \subseteq S$ of the states of the considered circuit is a *winning set* if all states in W satisfy the invariant P and, for all values of the primary inputs, the black boxes can ensure (by computing appropriate values) that the successor state is again in W.

A given incomplete sequential circuit is realizable if there is a winning set that includes the initial state of the circuit. This can be formulated as the DQBF (2.7). Similar to the combinational case, we have to take into account that the black boxes are typically not directly connected to the primary inputs, but to internal signals. This is done by restricting the requirement that the successor state is again a winning state to the case when the black box inputs are assigned consistently with the values computed by their input cones.

This formula (together with the corresponding result for incomplete combinational circuits, see Sect. 2.3.4) also shows that deciding the realizability of invariants for

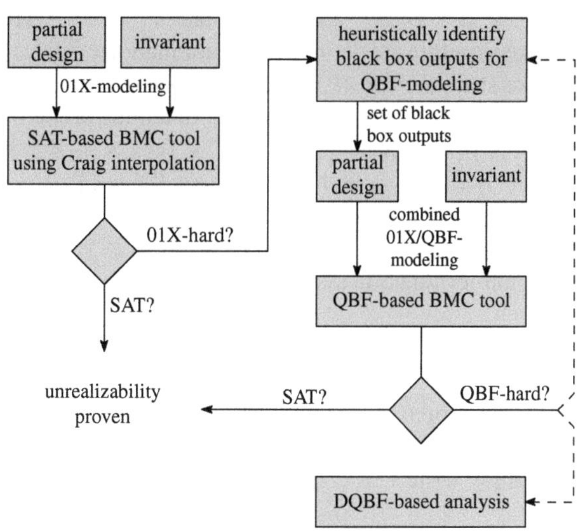

Fig. 2.6 Workflow

incomplete sequential circuits with combinational (or bounded-memory) black boxes is NEXPTIME-complete.

Figure 2.6 illustrates the complete verification flow for incomplete sequential circuits.

2.4.2 *Model Checking for Incomplete Designs*

In this section, we consider methods for incomplete designs which do not consider *bounded* model checking as in Sect. 2.4.1, but *symbolic* model checking using symbolic state set representations, i.e., state set representations based on binary decision diagrams (BDDs) [6] or and-inverter-graphs (AIGs). The approach generalizes symbolic model checking for complete designs [8] to (approximate) symbolic model checking for *incomplete* designs. In doing so, we can extend the consideration of invariant properties to the consideration of general CTL (computation tree logic) properties [9].

2.4.2.1 Symbolic Model Checking for Complete Designs

Symbolic model checking for complete designs [8] is applied to Kripke structures on the one hand, which may be derived from sequential circuits, and to a formula of a temporal logic (in our case: computation tree logic, CTL) on the other hand. A (complete) sequential circuit (such as considered in the section before) defines a Mealy automaton $M = (\mathbb{B}^{|\mathbf{s}|}, \mathbb{B}^{|\mathbf{x}|}, \mathbb{B}^{|\mathbf{y}|}, \delta, \lambda, \mathbf{s}^0)$ with state set $\mathbb{B}^{|\mathbf{s}|}$, the set of inputs $\mathbb{B}^{|\mathbf{x}|}$, the set of outputs $\mathbb{B}^{|\mathbf{y}|}$, transition function $\delta\colon \mathbb{B}^{|\mathbf{s}|} \times \mathbb{B}^{|\mathbf{x}|} \to \mathbb{B}^{|\mathbf{s}|}$, output function $\lambda\colon \mathbb{B}^{|\mathbf{s}|} \times \mathbb{B}^{|\mathbf{x}|} \to \mathbb{B}^{|\mathbf{y}|}$ and initial state $\mathbf{s}^0 \in \mathbb{B}^{|\mathbf{s}|}$.[5] The states of the corresponding Kripke structure are defined as a combination of states and inputs of M. The resulting Kripke structure for M is given by $struct(M) = (S, R, L)$ where $S = \mathbb{B}^{|\mathbf{s}|} \times \mathbb{B}^{|\mathbf{x}|}$, $R \subseteq S \times S, L : S \to V$. V is the set of atomic properties, i.e., $V = \{x_0, \ldots, x_{|\mathbf{x}|-1}\} \cup \{y_0, \ldots, y_{|\mathbf{y}|-1}\}$, $R = \{((\mathbf{s}, \mathbf{x}), (\mathbf{s}', \mathbf{x}')) \mid \mathbf{s}, \mathbf{s}' \in \mathbb{B}^{|\mathbf{s}|}, \mathbf{x}, \mathbf{x}' \in \mathbb{B}^{|\mathbf{x}|}, \delta(\mathbf{s}, \mathbf{x}) = \mathbf{s}'\}$, and $L((\mathbf{s}, '')) = \{x_i \mid \varepsilon_i = 1\} \cup \{y_i \mid \lambda_i(\mathbf{s}, '') = 1\}$.

Given a set V of atomic propositions, the syntax of CTL formulas is given by the following context-free grammar, where $v \in V$ is an atomic proposition and Φ the only non-termination symbol:

$$\Phi ::= v \mid \neg\Phi \mid (\Phi \vee \Phi) \mid EX\Phi \mid EG\Phi \mid E\Phi U\Phi.$$

We write $struct(M), s \models \varphi$ if φ is a CTL formula that is satisfied in state $s = (\mathbf{s}, \mathbf{x}) \in S$ of $struct(M)$. If it is clear from the context which Kripke structure is used, we simply write $s \models \varphi$ instead of $struct(M), s \models \varphi$. $\models$ is defined as follows:

---

[5]Here, we assume to have exactly one initial state; it is trivial to extend the following to sets of initial states.

(a) Fixed point iteration for *EG*

```
χ_EG(χ_X) {
  old := 1;
  new := χ_X;
  while (old ≠ new) {
    old := new;
    new := χ_X · χ_EX(old);
  }
  return new;
}
```

(b) Fixed point iteration for *EU*

```
χ_EU(χ_X, χ_Y) {
  old := 0;
  new := χ_Y;
  while (old ≠ new) {
    old := new;
    new := χ_Y + (χ_X · χ_EX(old));
  }
  return new;
}
```

Fig. 2.7 Fixed point iteration algorithms

$$
\begin{array}{lll}
s \models \varphi;\ \varphi \in V & \Longleftrightarrow & \varphi \in L(s) \\
s \models \neg\varphi & \Longleftrightarrow & s \not\models \varphi \\
s \models (\varphi_1 \vee \varphi_2) & \Longleftrightarrow & s \models \varphi_1 \text{ or } s \models \varphi_2 \\
s \models EX\varphi & \Longleftrightarrow & \exists s' \in S\colon\ R(s, s') \text{ and } s' \models \varphi \\
s \models EG\varphi & \Longleftrightarrow & \text{there is a path}(s_0, s_1, s_2, \ldots) \text{ with} \\
 & & s = s_0 \text{ and } \forall i \geq 0\colon\ (s_i, s_{i+1}) \in R \text{ and } s_i \models \varphi \\
s \models E\varphi_1 U \varphi_2 & \Longleftrightarrow & \text{there is a path } (s_0, s_1, s_2, \ldots) \text{ with} \\
 & & s = s_0 \text{ and } \forall i \geq 0\colon\ (s_i, s_{i+1}) \in R \text{ and there is} \\
 & & \text{a } j \text{ so that } s_j \models \varphi_2 \text{ and } \forall 0 \leq i < j\colon s_i \models \varphi_1
\end{array}
$$

Further CTL operations like $\wedge$, EF, AX, AU, AG, and AF can be expressed by using $\neg$, $\vee$, EX, EU, and EG [20].

In symbolic model checking, sets of states are represented by characteristic functions, which are in turn represented by BDDs. Let $Sat(\varphi)$ be the set of states of $struct(M)$ which satisfy formula φ and let $\chi_{Sat(\varphi)}$ be its characteristic function, then $\chi_{Sat(\varphi)}$ can be computed recursively based on the characteristic function $\chi_R(\mathbf{s}, \mathbf{x}, \mathbf{s}') := \prod_{i=0}^{|\mathbf{s}|-1} \left(\delta_i(\mathbf{s}, \mathbf{x}) \equiv s_i'\right)$ of the transition relation R:

$$
\begin{array}{ll}
\chi_{Sat(x_i)}(\mathbf{s}, \mathbf{x}) & := x_i \\
\chi_{Sat(y_i)}(\mathbf{s}, \mathbf{x}) & := \lambda_i(\mathbf{s}, \mathbf{x}) \\
\chi_{Sat(\neg\varphi)}(\mathbf{s}, \mathbf{x}) & := \overline{\chi_{Sat(\varphi)}}(\mathbf{s}, \mathbf{x}) \\
\chi_{Sat((\varphi_1 \vee \varphi_2))}(\mathbf{s}, \mathbf{x}) & := \chi_{Sat(\varphi_1)}(\mathbf{s}, \mathbf{x}) \vee \chi_{Sat(\varphi_2)}(\mathbf{s}, \mathbf{x}) \\
\chi_{Sat(EX\varphi)}(\mathbf{s}, \mathbf{x}) & := \chi_{EX}(\chi_{Sat(\varphi)})(\mathbf{s}, \mathbf{x}) \\
\chi_{Sat(EG\varphi)}(\mathbf{s}, \mathbf{x}) & := \chi_{EG}(\chi_{Sat(\varphi)})(\mathbf{s}, \mathbf{x}) \\
\chi_{Sat(E\varphi_1 U \varphi_2)}(\mathbf{s}, \mathbf{x}) & := \chi_{EU}(\chi_{Sat(\varphi_1)}, \chi_{Sat(\varphi_2)})(\mathbf{s}, \mathbf{x})
\end{array}
$$

with

$$\chi_{EX}(\chi_X)(\mathbf{s}, \mathbf{x}) := \exists \mathbf{s}' \exists \mathbf{x}' \Big(\chi_R(\mathbf{s}, \mathbf{x}, \mathbf{s}') \cdot \big(\chi_X|_{\substack{\mathbf{s} \leftarrow \mathbf{s}' \\ \mathbf{x} \leftarrow \mathbf{x}'}}\big)(\mathbf{s}', \mathbf{x}')\Big)$$

χ_{EG} and χ_{EU} can be evaluated by the fixed point iteration algorithms shown in Fig. 2.7.

A Mealy automaton satisfies a formula φ iff φ is satisfied in all the states of the corresponding Kripke structure which are derived from the initial state $\mathbf{s}^0$ of M:

$$M \models \varphi :\Longleftrightarrow \forall \mathbf{x} \in \mathbb{B}^{|\mathbf{x}|} : struct(M), (\mathbf{s}^0, \mathbf{x}) \models \varphi \quad \Longleftrightarrow \big(\forall \mathbf{x}(\chi_{Sat(\varphi)}|_{\mathbf{s}=\mathbf{s}^0})\big) = 1.$$

2.4.2.2 An Approximate Symbolic Model Checking Method for Incomplete Designs with Flexible Handling of Unknowns

(1) Flexible modeling of black box outputs in symbolic simulation

For symbolic CTL model checking of a given design, a symbolic representation of its output function λ and of its transition function δ are needed first. In order to generalize CTL model checking to *incomplete* designs, the potential effect of the black box outputs to the remaining design needs to be modeled in order to compute λ and δ. As in the sections before, we consider two different methods modeling black box outputs with differing accuracy: The first one (symbolic $(0, 1, X)$-simulation) is based on ternary $(0, 1, X)$ logic and the second one (symbolic Z_i-simulation) introduces for each output of a black box a separate variable Z_i. Similarly to Sect. 2.4.1.1, we consider a method for flexible modeling of different black box outputs by differing methods. This method will be applied later on for our approximate model checking algorithm.

For *symbolic* $(0, 1, X)$-simulation a new variable Z is introduced, which is used to model the unknown value X of the black box outputs. Now, for each output g_i of the incomplete design with primary input variables $x_1, \ldots, x_n$, a BDD representation of g_i is obtained by using a slightly modified version of symbolic simulation [7]. g_i depends on variables $x_1, \ldots, x_n$ and Z and has the following property:

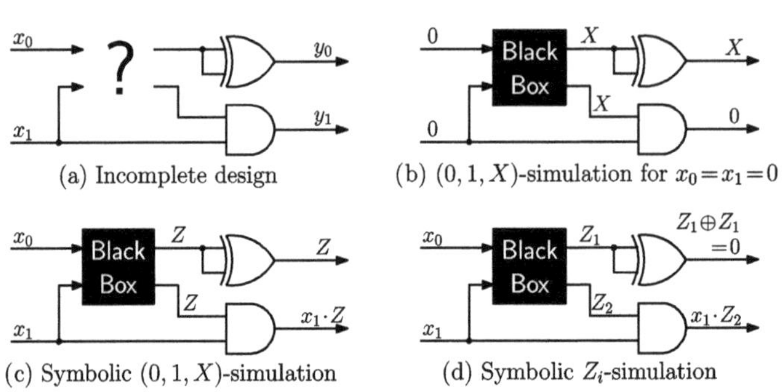

Fig. 2.8 Different methods to analyze an incomplete design

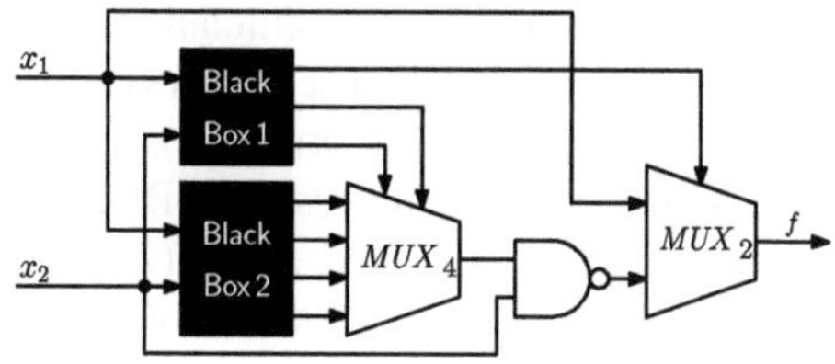

Fig. 2.9 An exemplary incomplete circuit

$$g_i|_{\substack{x_1=\varepsilon_1 \\ \dots \\ x_n=\varepsilon_n}} = \begin{cases} 1, & \text{if (0,1,}X\text{)-simulation with input } (\varepsilon_1, \dots, \varepsilon_n) \text{ produces } 1 \\ 0, & \text{if (0,1,}X\text{)-simulation with input } (\varepsilon_1, \dots, \varepsilon_n) \text{ produces } 0 \\ Z, & \text{if (0,1,}X\text{)-simulation with input } (\varepsilon_1, \dots, \varepsilon_n) \text{ produces } X \end{cases} \tag{2.8}$$

See Fig. 2.8c for an example.

To compute BDDs for the functions g_i by symbolic simulation the inputs of the circuit are associated with unique BDD variables as in a conventional symbolic simulation [7]. All output signals of black boxes are associated with the new variable Z. Now BDDs for the functions computed by the gates of the circuit are built in topological order treating the black box outputs (associated with variable Z) like inputs of the circuit. The gates of the circuit can then be processed in a manner similar to a conventional symbolic simulation.[6] When we process an and_2 (or_2) gate, we combine the BDDs for the two predecessor functions by a BDD *AND* (*OR*) operation as in the conventional symbolic simulation. For an inv gate we perform a *NOT* operation on the BDD of the predecessor function, now followed by a *compose* operation (see, i.e., [6]) which composes $\overline{Z}$ for Z (written as $g|_{Z \leftarrow \overline{Z}}$ for a composition of $\overline{Z}$ for Z in g).

It is easy to see that this (modified) symbolic simulation leads to BDD representations fulfilling property (2.8).

In symbolic Z_i-simulation, a new variable Z_i is introduced for each black box output instead of using the same variable Z for all black box outputs, and a (conventional) symbolic simulation is performed. Figure 2.8d shows an example for symbolic Z_i-simulation. (Note that—in contrast to symbolic $(0, 1, X)$-simulation in Fig. 2.8c—the first output can now be proven to be constant 0.)

Flexible Z/Z_i-representation of incomplete circuits combines the two methods and allows some black box outputs to be represented as in symbolic $(0, 1, X)$-simulation and some black box outputs as in symbolic Z_i-simulation. Such a flexible representation of an incomplete circuit is computed as follows:

For each output of the black boxes, which are to be handled as in symbolic $(0, 1, X)$-simulation, we use the variable Z to model the black box output, while for each output of the black boxes, which are to be handled by symbolic Z_i-simulation we

[6] Since all types of gates can be expressed using two-input and_2 gates, two-input or_2 gates and inv gates, we can assume w. l. o. g. that the gates have types and_2, or_2 or inv.

use a new Z_i variable. The simulation now considers the latter black box outputs as additional inputs and then performs symbolic (0, 1, X)-simulation (always replacing Z by $\overline{Z}$ when processing *inv* gates).

It is easy to see that the resulting BDD representation for the circuit outputs g_i with primary input variables $x_1, \ldots, x_n$, (Z_i-simulated) black box outputs $Z_1, \ldots, Z_m$ and the Z-variable as inputs have the following property:

$$g_i|_{\substack{x_1=\varepsilon_1 \\ \ldots \\ x_n=\varepsilon_n \\ Z_1=\eta_1 \\ \ldots \\ Z_m=\eta_m}} = \begin{cases} 1, \text{ if (0,1,}X\text{)-simulation with input} \\ \quad (\varepsilon_1, \ldots, \varepsilon_n, \eta_1, \ldots, \eta_m) \text{ produces } 1 \\ 0, \text{ if (0,1,}X\text{)-simulation with input} \\ \quad (\varepsilon_1, \ldots, \varepsilon_n, \eta_1, \ldots, \eta_m) \text{ produces } 0 \\ Z, \text{ if (0,1,}X\text{)-simulation with input} \\ \quad (\varepsilon_1, \ldots, \varepsilon_n, \eta_1, \ldots, \eta_m) \text{ produces } X. \end{cases}$$

Example 2.6 Figure 2.9 shows an example: If this circuit is simulated by using symbolic (0, 1, X)-simulation (meaning that Z is assigned to the outputs of both black box 1 and black box 2), a total number of 3 variables are needed (x_1, x_2, Z) and the resulting function for the output is $f_Z = Z$.

If the circuit is simulated by using symbolic Z_i-simulation instead (meaning that for each output of black box 1 and black box 2 a new Z_i variable is used), 9 variables are needed ($x_1, x_2, Z_1, \ldots, Z_7$), and the function for the output is $f_{Z_i} = \overline{Z}_1 x_1 + Z_1 \cdot \big(\overline{x}_2 + \neg(\overline{Z}_2 \overline{Z}_3 Z_4 + Z_2 \overline{Z}_3 Z_5 + \overline{Z}_2 Z_3 Z_6 + Z_2 Z_3 Z_7)\big)$ (when variables $Z_1, \ldots Z_7$ are assigned top down to the black box outputs appearing in Fig. 2.9).

When using the flexible Z/Z_i-method for modeling black box outputs, assigning Z to all outputs of black box 2, but different Z_i's to the outputs of black box 1, e.g., we end up using six variables ($x_1, x_2, Z, Z_1, Z_2, Z_3$) and obtain the function $f_{\text{flex}} = \overline{Z}_1 x_1 + Z_1 \cdot (\overline{x}_2 + Z)$.

So, the flexible method generates an output function that is obviously less complicated than the result of symbolic Z_i-simulation, yet contains more information than the result of symbolic (0, 1, X)-simulation. To give an example, for $x_1 = 1$ and $x_2 = 0$, the output can be proven to be 1 using the flexible method, while it is not possible to obtain this information from symbolic (0, 1, X)-simulation.

(2) Basic principle of symbolic model checking for incomplete designs

We start by describing the basic principle. Symbolic model checking for complete designs computes the set *Sat*(φ) of all states satisfying a CTL formula φ and then checks whether all initial states are included in this set. If so, the circuit satisfies φ.

The situation becomes more complex if we consider incomplete circuits, since for each replacement of the black boxes we may have different state sets satisfying φ. In contrast to conventional model checking, we will consider two sets instead of *Sat*(φ). These two sets result from different replacements of the black boxes in incomplete circuits:

Definition 2.5 Let I be an incomplete circuit and let r be a replacement of the black boxes in I by (combinational or sequential) circuits which transforms I into a complete circuit I^r. Let $\mathscr{R}$ be the set of all possible replacements of this kind. For a replacement $r \in \mathscr{R}$ let $Sat^r(\varphi)$ be the set of states of I^r satisfying φ. For the case that the replacement r contains *sequential* circuits, we define a set $Sat^r_E(\varphi)$ which results from $Sat^r(\varphi)$ by existentially quantifying the state bits corresponding to memory elements inside the black box replacements, and the set $Sat^r_A(\varphi)$ which results by universally quantifying these state bits.

- $Sat^{\text{exact}}_E(\varphi)$ is defined by $Sat^{\text{exact}}_E(\varphi) := \bigcup_{r\in\mathscr{R}} Sat^r_E(\varphi)$. We define that a state in $Sat^{\text{exact}}_E(\varphi)$ *possibly* satisfies the property φ.
- $Sat^{\text{exact}}_A(\varphi)$ is defined by $Sat^{\text{exact}}_A(\varphi) := \bigcap_{r\in\mathscr{R}} Sat^r_A(\varphi)$. We define that a state in $Sat^{\text{exact}}_A(\varphi)$ *definitely* satisfies the property φ.

$Sat^{\text{exact}}_E(\varphi)$ contains all states, for which *there is* at least one black box replacement (together with some initialization for the memory elements inside these replacements) so that φ is satisfied. In order to compute $Sat^{\text{exact}}_E(\varphi)$, we could *conceptually* consider all possible replacements $r \in \mathscr{R}$ of the black boxes, compute $Sat^r(\varphi)$ for each such replacement by conventional model checking, perform an existential quantification for the state bits of memory elements inside the black boxes, and determine $Sat^{\text{exact}}_E(\varphi)$ as the union of all sets $Sat^r_E(\varphi)$. (Of course, this approach is not feasible, since the set $\mathscr{R}$ may be large and is not even finite, if we allow replacements by sequential circuits).

In a similar manner, $Sat^{\text{exact}}_A(\varphi)$ contains all states, for which φ is satisfied for *all* possible black box replacements (regardless of the initial values for the memory elements inside black boxes).

Given $Sat^{\text{exact}}_E(\varphi)$ and $Sat^{\text{exact}}_A(\varphi)$, it is easy to prove validity and to falsify realizability for the incomplete circuit:

Lemma 2.1 *If all initial states are included in $Sat^{exact}_A(\varphi)$, φ is satisfied for all replacements of the black boxes ("φ is valid").*

If there is at least one initial state not belonging to $Sat^{exact}_E(\varphi)$, there is no replacement of the black boxes so that φ is satisfied for the resulting complete circuit ("φ is not realizable").

Proof If all initial states are included in $Sat^{\text{exact}}_A(\varphi)$, then for each replacement $r \in \mathscr{R}$ all initial states are included in $Sat^r_A(\varphi)$. This in turn means that all extensions of initial states by arbitrary initial values for memory elements inside the black boxes are included in $Sat^r(\varphi)$, i.e., φ is valid.

If there is at least one initial state $\mathbf{s}^0$ outside of $Sat^{\text{exact}}_E(\varphi)$, then $\mathbf{s}^0$ is not included in $Sat^r_E(\varphi)$ for all replacements $r \in \mathscr{R}$ of the black boxes. Thus, all extensions of $\mathbf{s}^0$ by initial values for memory elements inside the black boxes are not included in $Sat^r(\varphi)$. Since it is not possible to choose a replacement for the black boxes (together with some initialization to the memory elements of the black boxes) such that the initial state $\mathbf{s}^0$ satisfies φ, φ is not realizable.

Just as black boxes in incomplete circuits lead to states definitely satisfying φ on the one hand and states possibly satisfying φ on the other hand, there are also two types of transitions between states in an incomplete circuit:

- There are transitions which exist independently from the replacement of the black boxes, i.e., for all possible replacements of the black boxes (we will call them 'fixed transitions') and
- there are transitions which may or may not exist in a complete version of the design—depending on the implementation for the black boxes (we will call them 'possible transitions').

Formally, fixed and possible transitions are defined as follows:

Definition 2.6 Let I be an incomplete circuit with state set $\mathbb{B}^{|\mathbf{s}|}$ and set of inputs $\mathbb{B}^{|\mathbf{x}|}$. Let $r \in \mathscr{R}$ be a replacement of the black boxes leading to a complete circuit I^r and let $\mathbb{B}^{|\mathbf{sr}|}$ be the state space formed by $|\mathbf{sr}|$ state bits inside the black box replacements.

- The incomplete circuit has a *fixed* transition from state $(\mathbf{s}, \mathbf{x}) \in \mathbb{B}^{|\mathbf{s}|} \times \mathbb{B}^{|\mathbf{x}|}$ to state $(\mathbf{s}', \mathbf{x}')\mathbb{B}^{|\mathbf{s}|} \times \mathbb{B}^{|\mathbf{x}|}$, if for each replacement $r \in \mathscr{R}$ and all $\mathbf{sr}, \mathbf{sr}' \in \mathbb{B}^{|\mathbf{sr}|}$, there is a transition from $(\mathbf{s}, \mathbf{sr}, \mathbf{x})$ to $(\mathbf{s}', \mathbf{sr}', \mathbf{x}')$ in the Mealy automaton corresponding to I^r.
- The incomplete circuit has a *possible* transition from state $(\mathbf{s}, \mathbf{x}) \in \mathbb{B}^{|\mathbf{s}|} \times \mathbb{B}^{|\mathbf{x}|}$ to state $(\mathbf{s}', \mathbf{x}')\mathbb{B}^{|\mathbf{s}|} \times \mathbb{B}^{|\mathbf{x}|}$, if there is a replacement $r \in \mathscr{R}$ and there are $\mathbf{sr}, \mathbf{sr}' \in \mathbb{B}^{|\mathbf{sr}|}$, such that there is a transition from $(\mathbf{s}, \mathbf{sr}, \mathbf{x})$ to $(\mathbf{s}', \mathbf{sr}', \mathbf{x}')$ in the Mealy automaton corresponding to I^r.

Fixed and possible transitions can be used in order to compute states that possibly or definitely satisfy a property φ.

(3) Approximations

For reasons of efficiency, we do not compute the exact sets of fixed and possible transitions and we do not compute the exact sets $Sat_E^{\text{exact}}(\varphi)$ and $Sat_A^{\text{exact}}(\varphi)$.

Instead of $Sat_E^{\text{exact}}(\varphi)$ and $Sat_A^{\text{exact}}(\varphi)$ we compute *approximations* $Sat_E(\varphi)$ and $Sat_A(\varphi)$ of these sets. To be more precise, we will compute over-approximations $Sat_E(\varphi) \supseteq Sat_E^{\text{exact}}(\varphi)$ of $Sat_E^{\text{exact}}(\varphi)$ and under-approximations $Sat_A(\varphi) \subseteq Sat_A^{\text{exact}}(\varphi)$ of $Sat_A^{\text{exact}}(\varphi)$.

Because of $Sat_E(\varphi) \supseteq Sat_E^{\text{exact}}(\varphi) \supseteq Sat_E^r(\varphi) \supseteq Sat^r(\varphi)$ for arbitrary replacements r of the black boxes with arbitrary initialization of memory elements inside black boxes we can also guarantee for $Sat_E(\varphi)$ that φ is not realizable if some initial state is not included in $Sat_E(\varphi)$. Analogously we can guarantee that φ is valid, if all initial states are included in $Sat_A(\varphi)$ (since $Sat_A(\varphi) \subseteq Sat_A^{\text{exact}}(\varphi) \subseteq Sat_A^r(\varphi) \subseteq Sat^r(\varphi)$).

This argumentation results in the following lemma:

Lemma 2.2 *Let $Sat_A(\varphi)$ be an under-approximation of $Sat_A^{exact}(\varphi)$, $Sat_E(\varphi)$ an over-approximation of $Sat_E^{exact}(\varphi)$. If all initial states are included in $Sat_A(\varphi)$, then φ is valid. If there is an initial state that is not included in $Sat_E(\varphi)$, then φ is not realizable.*

Approximations $Sat_E(\varphi)$ and $Sat_A(\varphi)$ will be computed based on an approximate transition relation and on approximate output functions for the corresponding Mealy automaton M. Approximations of transition function δ and output function λ are computed using symbolic Z/Z_i-simulations as defined above.

We start with the computation of two approximations of the sets of states in which λ_i is true, i.e., we start with the computation of under-approximations $Sat_A(y_i)$ and overapproximations $Sat_E(y_i)$. Under-approximations $Sat_A(\varphi)$ and over-approximations $Sat_E(\varphi)$ for arbitrary CTL formulas φ will be considered later on in this section.

For an incomplete circuit, let there be a number of black boxes with outputs modeled by Z and some other black boxes with outputs modeled by Z_i's. We apply symbolic Z/Z_i-simulation for computing the transition functions and the output functions. Thus, we introduce new variables Z and $\mathbf{Z}_l = (Z_{l,1}, Z_{l,2}, \ldots)$. The symbolic symbolic Z/Z_i-simulation now provides symbolic representations of the output functions $\lambda_i(\mathbf{s}, \mathbf{x}, Z, \mathbf{Z}_l)$ and transition functions $\delta_j(\mathbf{s}, \mathbf{x}, Z, \mathbf{Z}_l)$.

In standard model checking for complete designs, an atomic property y_i is satisfied for a state $(\mathbf{s}^{\text{fix}}, \mathbf{x}^{\text{fix}}) \in \mathbb{B}^{|\mathbf{s}| \times |\mathbf{x}|}$ if $\lambda_i|_{\mathbf{s}=\mathbf{s}^{\text{fix}}, \mathbf{x}=\mathbf{x}^{\text{fix}}} = 1$.

Here we include a state $(\mathbf{s}^{\text{fix}}, \mathbf{x}^{\text{fix}})$ into $Sat_A(y_i)$, if λ_i is 1 for $(\mathbf{s}^{\text{fix}}, \mathbf{x}^{\text{fix}})$ assigned to $(\mathbf{s}, \mathbf{x})$ and *all* possible assignments to Z and $\mathbf{Z}_l$. We include $(\mathbf{s}^{\text{fix}}, \mathbf{x}^{\text{fix}})$ into $Sat_E(y_i)$, if λ_i is 1 for $(\mathbf{s}^{\text{fix}}, \mathbf{x}^{\text{fix}})$ assigned to $(\mathbf{s}, \mathbf{x})$ and some assignment to Z and $\mathbf{Z}_l$. Thus we define the characteristic functions of $Sat_A(y_i)$ and $Sat_E(y_i)$ as follows:

Definition 2.7

$$\chi_{Sat_A(y_i)}(\mathbf{s}, \mathbf{x}) := \forall Z \forall \mathbf{Z}_l \big(\lambda_i(\mathbf{s}, \mathbf{x}, Z, \mathbf{Z}_l)\big) \tag{2.9}$$

$$\chi_{Sat_E(y_i)}(\mathbf{s}, \mathbf{x}) := \exists Z \exists \mathbf{Z}_l \big(\lambda_i(\mathbf{s}, \mathbf{x}, Z, \mathbf{Z}_l)\big) \tag{2.10}$$

Lemma 2.3 *For $Sat_A(y_i)$ and for $Sat_E(y_i)$ as defined in Definition 2.7:*

$$Sat_A(y_i) \subseteq Sat_A^{exact}(y_i), \quad Sat_E^{exact}(y_i) \subseteq Sat_E(y_i).$$

Proof If $\lambda_i|_{\mathbf{s}=\mathbf{s}^{\text{fix}}, \mathbf{x}=\mathbf{x}^{\text{fix}}} = 1$ for some state $(\mathbf{s}^{\text{fix}}, \mathbf{x}^{\text{fix}}) \in \mathbb{B}^{|\mathbf{s}| \times |\mathbf{x}|}$, then we know that λ_i is 1 in this state independently from the replacement of the black boxes, so $(\mathbf{s}^{\text{fix}}, \mathbf{x}^{\text{fix}})$ can be included into $Sat_A(y_i)$ and $Sat_E(y_i)$.

If $\lambda_i|_{\mathbf{s}=\mathbf{s}^{\text{fix}}, \mathbf{x}=\mathbf{x}^{\text{fix}}} = 0$, then the output λ_i is 0 in this state independently from the replacement of the black boxes, so we can include $(\mathbf{s}^{\text{fix}}, \mathbf{x}^{\text{fix}})$ neither into $Sat_A(y_i)$ nor into $Sat_E(y_i)$.

In any other case, the value of y_i is unknown in this state and thus we can include $(\mathbf{s}^{\text{fix}}, \mathbf{x}^{\text{fix}})$ into $Sat_E(y_i)$, but not into $Sat_A(y_i)$.

By Eq. (2.9) only states $(\mathbf{s}^{\text{fix}}, \mathbf{x}^{\text{fix}})$ with $\lambda_i|_{\mathbf{s}=\mathbf{s}^{\text{fix}}, \mathbf{x}=\mathbf{x}^{\text{fix}}} = 1$ are included in $Sat_A(y_i)$, and by Eq. (2.10) all states $(\mathbf{s}^{\text{fix}}, \mathbf{x}^{\text{fix}})$ with $\lambda_i|_{\mathbf{s}=\mathbf{s}^{\text{fix}}, \mathbf{x}=\mathbf{x}^{\text{fix}}} \neq 0$ are included in $Sat_E(y_i)$.

Now the computation of $Sat_A(\varphi)$ and $Sat_E(\varphi)$ is performed based on fixed and possible transitions. Here we work with approximations, too: We compute an over-approximation of the possible transitions, represented by the characteristic function

$\chi_{R_E}(\mathbf{s}, \mathbf{x}, \mathbf{s}')$. (An under-approximation $\chi_{R_A}(\mathbf{s}, \mathbf{x}, \mathbf{s}')$ containing at most the fixed transitions could be computed as well, however it turns out that it is not really needed for our algorithm.)

We define our over-approximation of the possible transitions as follows:

Definition 2.8

$$\chi_{R_E}(\mathbf{s}, \mathbf{x}, \mathbf{s}') := \exists \mathbf{Z}_l \Big(\prod_{i=0}^{|\mathbf{s}|-1} \exists Z \big(\delta_i(\mathbf{s}, \mathbf{x}, Z, \mathbf{Z}_l) \equiv s_i'\big)\Big). \qquad (2.11)$$

The following lemma states that χ_{R_E} over-approximates the possible transitions:

Lemma 2.4 *If $\chi_{R_E}(\mathbf{s}^{fix}, \mathbf{x}^{fix}, \mathbf{s}'^{fix}) = 0$, then there is no possible transition from $(\mathbf{s}^{fix}, \mathbf{x}^{fix})$ to $(\mathbf{s}'^{fix}, \mathbf{x}'^{fix})$ (for an arbitrary next input $\mathbf{x}'^{fix}$).*

Proof If $\chi_{R_E}(\mathbf{s}^{\text{fix}}, \mathbf{x}^{\text{fix}}, \mathbf{s}'^{\text{fix}}) = 0$, then $\forall \mathbf{Z}_l \Big(\bigvee_{i=0}^{|\mathbf{s}|-1} \forall Z \big(\delta_i(\mathbf{s}^{\text{fix}}, \mathbf{x}^{\text{fix}}, Z, \mathbf{Z}_l) \not\equiv s_i'^{\text{fix}}\big)\Big) = 1$. This means that for an arbitrary fixed output $\mathbf{Z}_l^{\text{fix}}$ of the black boxes modeled by Z_i's there is an $0 \le i \le |\mathbf{s}|-1$ with $\forall Z \big(\delta_i(\mathbf{s}^{\text{fix}}, \mathbf{x}^{\text{fix}}, Z, \mathbf{Z}_l^{\text{fix}}) \not\equiv s_i'^{\text{fix}}\big) = 1 (\star)$. According to the symbolic Z-simulation, $\delta_i(\mathbf{s}^{\text{fix}}, \mathbf{x}^{\text{fix}}, Z, \mathbf{Z}_l^{\text{fix}}) = Z$, $\delta_i(\mathbf{s}^{\text{fix}}, \mathbf{x}^{\text{fix}}, Z, \mathbf{Z}_l^{\text{fix}}) = s_i'^{\text{fix}}$, or $\delta_i(\mathbf{s}^{\text{fix}}, \mathbf{x}^{\text{fix}}, Z, \mathbf{Z}_l^{\text{fix}}) = \neg s_i'^{\text{fix}}$. In the two former cases, $(\star)$ would not hold, thus we have $\delta_i(\mathbf{s}^{\text{fix}}, \mathbf{x}^{\text{fix}}, Z, \mathbf{Z}_l^{\text{fix}}) = \neg s_i'^{\text{fix}}$, i.e., the output value of δ_i differs from $s_i'^{\text{fix}}$ independently from the behavior of the Z-modeled black boxes.

Altogether we can conclude that the output value of δ for input $(\mathbf{s}^{\text{fix}}, \mathbf{x}^{\text{fix}})$ differs from $\mathbf{s}'^{\text{fix}}$ independently from the values at the outputs of black boxes, i.e., there cannot be a possible transition from $(\mathbf{s}^{\text{fix}}, \mathbf{x}^{\text{fix}})$ to $(\mathbf{s}'^{\text{fix}}, \mathbf{x}'^{\text{fix}})$.

Remark 2.1 Extending Definitions 2.5 and 2.6 with respect to our approximations, we will denote in the following not only the states in $Sat_E^{\text{exact}}(\varphi)$, but also the states in $Sat_E(\varphi)$ by 'states *possibly* satisfying φ.' Similarly, we characterize all transitions described by χ_{R_E} as 'possible transitions.'

Based on χ_{R_E}, $Sat_A(y_i)$ and $Sat_E(y_i)$, it is possible to define rules how arbitrary CTL formulas can be recursively evaluated. We show here how to compute sets $Sat_A(\cdot)$ and $Sat_E(\cdot)$ for CTL formulas $\neg\psi$, $(\psi_1 \vee \psi_2)$, $EG\psi$, and $E\psi_1 U \psi_2$.

For each state $(\mathbf{s}, \mathbf{x})$, $\chi_{R_E}(\mathbf{s}, \mathbf{x}, \mathbf{s}')$ gives us the set of $\mathbf{s}'$values the possible successors can have. Each of these different $\mathbf{s}'$values represents a *set* $S_{\mathbf{s}'} := \{(\mathbf{s}', \mathbf{x}') | \mathbf{x}' \in \mathbb{B}^{|\mathbf{x}|}\}$ of possible successor states sharing this $\mathbf{s}'$ value (yet with arbitrary value of $\mathbf{x}'$). So, if for a state $(\mathbf{s}, \mathbf{x})$ one of the states in one of these possible successor sets $S_{\mathbf{s}'}$ possibly satisfies ψ (i.e., is in $Sat_E(\psi)$), the current state possibly satisfies $EX\psi$ and can thus be included in $Sat_E(EX\psi)$. Figure 2.10 illustrates the sets.

Definition 2.9 *$Sat_E(EX\psi)$ is the set of states for which there is a possible successor that is in $Sat_E(\psi)$. This is represented by:*

$$\chi_{Sat_E(EX\psi)}(\mathbf{s}, \mathbf{x}) := \exists \mathbf{s}'\Big(\chi_{R_E}(\mathbf{s}, \mathbf{x}, \mathbf{s}') \cdot \exists \mathbf{x}'\big(\chi_{Sat_E(\psi)}|_{\substack{\mathbf{q}\leftarrow\mathbf{q}' \\ \mathbf{x}\leftarrow\mathbf{x}'}}\big)(\mathbf{s}', \mathbf{x}')\Big).$$

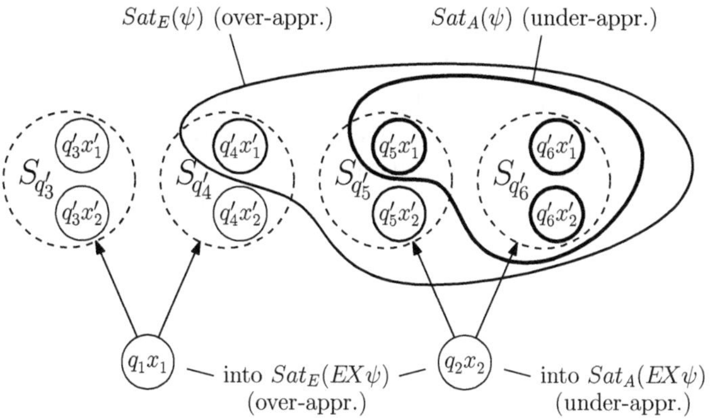

Fig. 2.10 Evaluation of $Sat_A(EX\psi)$, $Sat_E(EX\psi)$

On the other hand, if in each set $S_{\mathbf{s}'}$ of possible successors of $(\mathbf{s}, \mathbf{x})$ there is at least one state that definitely satisfies ψ (i.e., is in $Sat_A(\psi)$), then for each black box implementation at least one successor of state $(\mathbf{s}, \mathbf{x})$ satisfies ψ and thus, the current state $(\mathbf{s}, \mathbf{x})$ definitely satisfies $EX\psi$ and can be included in $Sat_A(EX\psi)$. Again, Fig. 2.10 illustrates the sets.

Definition 2.10 $Sat_A(EX\psi)$ is the set of states for which in each set $S_{\mathbf{s}'}$ of possible successors there is at least one state that is in $Sat_A(\psi)$. This is represented by:

$$\chi_{Sat_A(EX\psi)}(\mathbf{s}, \mathbf{x}) := \forall \mathbf{s}'\Big(\chi_{R_E}(\mathbf{s}, \mathbf{x}, \mathbf{s}') \rightarrow \exists \mathbf{x}'\big(\chi_{Sat_A(\psi)}|_{\substack{\mathbf{q}\leftarrow\mathbf{q}'\\ \mathbf{x}\leftarrow\mathbf{x}'}}\big)(\mathbf{s}', \mathbf{x}')\Big).$$

Lemma 2.5 *Let $Sat_A(\psi)$ be an under-approximation of $Sat_A^{exact}(\psi)$ and $Sat_E(\psi)$ be an over-approximation of $Sat_E^{exact}(\psi)$. For $Sat_A(EX\psi)$ as defined in Definition 2.10 and for $Sat_E(EX\psi)$ as defined in Definition 2.9, the following holds:*

$$Sat_A(EX\psi) \subseteq Sat_A^{exact}(EX\psi), \quad Sat_E^{exact}(EX\psi) \subseteq Sat_E(EX\psi).$$

The proof of this lemma follows from the considerations given above the definitions.

Negation can be defined as follows: Since $Sat_E(\psi)$ is an over-approximation of all states in which ψ *may be* satisfied for some black box replacement, we do know that for an arbitrary state in $\mathbb{B}^{|\mathbf{s}|} \times \mathbb{B}^{|\mathbf{x}|} \setminus Sat_E(\psi)$ there is no black box replacement so that ψ is satisfied in this state or, equivalently, $\neg\psi$ is definitely satisfied in this state for all black box replacements. This means that we can use $\mathbb{B}^{|\mathbf{s}|} \times \mathbb{B}^{|\mathbf{x}|} \setminus Sat_E(\psi)$ as an under-approximation $Sat_A(\neg\psi)$. Since an analogous argument holds for $Sat_A(\psi)$ and $Sat_E(\neg\psi)$ we define

Definition 2.11 $\chi_{Sat_A(\neg\psi)}(\mathbf{s}, \mathbf{x}) := \overline{\chi_{Sat_E(\psi)}(\mathbf{s}, \mathbf{x})}$ and $\chi_{Sat_E(\neg\psi)}(\mathbf{s}, \mathbf{x}) := \overline{\chi_{Sat_A(\psi)}(\mathbf{s}, \mathbf{x})}$.

$Sat_A(\psi_1 \vee \psi_2)$ and $Sat_E(\psi_1 \vee \psi_2)$ are computed as usual in model checking for complete designs.

Definition 2.12 $\chi_{Sat_A(\psi_1 \vee \psi_2)}(\mathbf{s}, \mathbf{x}) := (\chi_{Sat_A(\psi_1)} \vee \chi_{Sat_A(\psi_2)})(\mathbf{s}, \mathbf{x})$ and $\chi_{Sat_E(\psi_1 \vee \psi_2)}(\mathbf{s}, \mathbf{x}) := (\chi_{Sat_E(\psi_1)} \vee \chi_{Sat_E(\psi_2)})(\mathbf{s}, \mathbf{x})$.

Finally, we define $\varphi = EG\psi$ and $\varphi = E\psi_1 U \psi_2$ to be evaluated by their standard fixed point iterations (see Fig. 2.7) based on the evaluation of EX defined above (two separate fixed point iterations for Sat_A and Sat_E). We do not need to define more CTL operations, since other CTL operations can be expressed using the operations discussed so far.

Lemma 2.6 *For the sets $Sat_A(\neg\psi)$, $Sat_E(\neg\psi)$, $Sat_A(\psi_1 \vee \psi_2)$, $Sat_E(\psi_1 \vee \psi_2)$, $Sat_A(EG\psi)$, $Sat_E(EG\psi)$, $Sat_A(E\psi_1 U \psi_2)$, and $Sat_E(E\psi_1 U \psi_2)$ as defined above, the following holds:*

$$\begin{aligned}
Sat_A(\neg\psi) &\subseteq Sat_A^{exact}(\neg\psi), \\
Sat_E^{exact}(\neg\psi) &\subseteq Sat_E(\neg\psi), \\
Sat_A(\psi_1 \vee \psi_2) &\subseteq Sat_A^{exact}(\psi_1 \vee \psi_2), \\
Sat_E^{exact}(\psi_1 \vee \psi_2) &\subseteq Sat_E(\psi_1 \vee \psi_2), \\
Sat_A(EG\psi) &\subseteq Sat_A^{exact}(EG\psi), \\
Sat_E^{exact}(EG\psi) &\subseteq Sat_E(EG\psi), \\
Sat_A(E\psi_1 U \psi_2) &\subseteq Sat_A^{exact}(E\psi_1 U \psi_2), \\
Sat_E^{exact}(E\psi_1 U \psi_2) &\subseteq Sat_E(E\psi_1 U \psi_2).
\end{aligned}$$

Theorem 2.1 *The result of the recursive computation can be evaluated as follows*[7]*:*

$$\begin{aligned}
\left(\forall \mathbf{x}(\chi_{Sat_A(\varphi)}|_{\mathbf{s}=\mathbf{s}^0})\right) = 1 \quad &\Longrightarrow \varphi \text{ is valid} \\
\left(\exists \mathbf{x}(\overline{\chi_{Sat_E(\varphi)}}|_{\mathbf{s}=\mathbf{s}^0})\right) = 1 &\Longrightarrow \varphi \text{ is not realizable.}
\end{aligned}$$

Proof The proof follows directly from Lemmas 2.2, 2.3, 2.5, and 2.6.

(4) Including Z_i-variables into the state space

A further improvement on the accuracy of the two approximated sets considered above can be obtained by including Z_i-variables assigned to black box outputs into the state space.

As a motivation for this, consider the simple CTL formula $EF(y \wedge \neg y)$ for a design in which a black box output is directly connected to the primary output y. In every state, $(\mathbf{s}, \mathbf{x})$ both y and $\neg y$ are *possibly* satisfied (depending on the black box implementation), but they are not *definitely* satisfied. Thus, the method described so far computes the result that $y \wedge \neg y$ is possibly satisfied in every state $(\mathbf{s}, \mathbf{x})$, but not definitely, and the same result holds for $EF(y \wedge \neg y)$. For this reason, the method is neither able to prove validity nor to falsify realizability for the given incomplete design and the given formula.

However, it is clear that there will be no point in time during the computation where y is simultaneously true and false. Problems of this kind can be solved if

[7] Remember that $\mathbf{s}^0$ is the initial state of the circuit.

we include Z_i-variables assigned to black box outputs into the states of the Kripke structure. In this way, the according black box output values Z_i are constant within each single state and therefore in our example y has a fixed value for each state.

Note that it is not always necessary to include *all* Z_i's into the state space; this provides another possibility of flexibly processing the unknowns at this point, which can be used as a tradeoff between efficiency and accuracy.

Let $\mathbf{Z}_o$ be the Z_i-simulated black box outputs that are included into the state space and let $\mathbf{Z}_l$ be the Z_i-simulated black box outputs that are not included. Then the values of $\mathbf{Z}_o$ are constant within each single state, while the values of $\mathbf{Z}_l$ are arbitrary as they were before.

Both the output function $\lambda(\mathbf{s}, \mathbf{x}, Z, \mathbf{Z}_l, \mathbf{Z}_o)$ and the transition function $\delta(\mathbf{s}, \mathbf{x}, Z, \mathbf{Z}_l, \mathbf{Z}_o)$ can be computed by using symbolic Z/Z_i-simulation, where for symbolic simulation it is not necessary to distinguish between $\mathbf{Z}_l$ and $\mathbf{Z}_o$.

The computation of sets $Sat_A(\cdot)$ and $Sat_E(\cdot)$ is performed in a manner similar to the previous section. We start with the sets of states definitely or possibly satisfying the atomic CTL formula y_i:

We include a state $(\mathbf{s}^{\text{fix}}, \mathbf{x}^{\text{fix}}, \mathbf{Z}_o^{\text{fix}}) \in \mathbb{B}^{|\mathbf{s}|\times|\mathbf{x}|\times|\mathbf{Z}_o|}$ into $Sat_A(y_i)$ (and $Sat_E(y_i)$), if λ_i is 1 for $(\mathbf{s}^{\text{fix}}, \mathbf{x}^{\text{fix}}, \mathbf{Z}_o^{\text{fix}})$ assigned to $(\mathbf{s}, \mathbf{x}, \mathbf{Z}_o)$ and *all* possible assignments to Z and $\mathbf{Z}_l$, since in this case λ_i is 1 in this state independently from the replacement of the black boxes.

We include $(\mathbf{s}^{\text{fix}}, \mathbf{x}^{\text{fix}}, \mathbf{Z}_o^{\text{fix}}) \in \mathbb{B}^{|\mathbf{s}|\times|\mathbf{x}|\times|\mathbf{Z}_o|}$ into $Sat_E(y_i)$, if λ_i is 1 for $(\mathbf{s}^{\text{fix}}, \mathbf{x}^{\text{fix}}, \mathbf{Z}_o^{\text{fix}})$ assigned to $(\mathbf{s}, \mathbf{x}, \mathbf{Z}_o)$ and *some* assignment to Z and $\mathbf{Z}_l$. However, if λ_i is 0 for $(\mathbf{s}^{\text{fix}}, \mathbf{x}^{\text{fix}}, \mathbf{Z}_o^{\text{fix}})$ assigned to $(\mathbf{s}, \mathbf{x}, \mathbf{Z}_o)$ and *all* assignments to Z and $\mathbf{Z}_l$, we include $(\mathbf{s}^{\text{fix}}, \mathbf{x}^{\text{fix}}, \mathbf{Z}_o^{\text{fix}})$ neither into $Sat_A(y_i)$ nor $Sat_E(y_i)$, since then the output λ_i is 0 in this state independently from the replacement of the black boxes. Altogether we define the characteristic functions of $Sat_A(y_i)$ and $Sat_E(y_i)$ as

Definition 2.13

$$\chi_{Sat_A(y_i)}(\mathbf{s}, \mathbf{x}, \mathbf{Z}_o) := \forall Z \forall \mathbf{Z}_l \big(\lambda_i(\mathbf{s}, \mathbf{x}, Z, \mathbf{Z}_l, \mathbf{Z}_o)\big)$$
$$\chi_{Sat_E(y_i)}(\mathbf{s}, \mathbf{x}, \mathbf{Z}_o) := \exists Z \exists \mathbf{Z}_l \big(\lambda_i(\mathbf{s}, \mathbf{x}, Z, \mathbf{Z}_l, \mathbf{Z}_o)\big).$$

Lemma 2.7 *For $Sat_A(y_i)$ and for $Sat_E(y_i)$ as defined in Definition 2.13:*

$$\forall \mathbf{Z}_o \chi_{Sat_A(y_i)} \le \chi_{Sat_A^{exact}(y_i)}, \quad \chi_{Sat_E^{exact}(y_i)} \le \exists \mathbf{Z}_o \chi_{Sat_E(y_i)}.$$

The lemma follows from the considerations given above the definition.

Analogously, we define an over-approximation χ_{R_E} for the characteristic function of possible transitions:

Definition 2.14

$$\chi_{R_E}(\mathbf{s}, \mathbf{x}, \mathbf{Z}_o, \mathbf{s}') := \left(\exists \mathbf{Z}_l \prod_{i=0}^{|\mathbf{s}|-1} \exists Z \big(\delta_i(\mathbf{s}, \mathbf{x}, Z, \mathbf{Z}_l, \mathbf{Z}_o) \equiv s_i'\big)\right).$$

Based on $Sat_A(y_i)$, $Sat_E(y_i)$ and χ_{R_E}, the sets $Sat_A(EX\psi)$ and $Sat_E(EX\psi)$ can be computed by arguments similar to the previous section. The main difference is that we have to handle the additional variables $\mathbf{Z}_o$ in the state space correctly:

$\chi_{R_E}(\mathbf{s}, \mathbf{x}, \mathbf{Z}_o, \mathbf{s}')$ gives us the $\mathbf{s}'$ values of possible successors of a state $(\mathbf{s}, \mathbf{x}, \mathbf{Z}_o)$. Now each of these different $\mathbf{s}'$ values represents a set $S_{\mathbf{s}'} := \{(\mathbf{s}', \mathbf{x}', \mathbf{Z}'_o) \mid \mathbf{x}' \in \mathbb{B}^{|\mathbf{x}|}, \mathbf{Z}'_o \in \mathbb{B}^{|\mathbf{Z}_o|}\}$ of possible successor states sharing this $\mathbf{s}'$ value. For $Sat_E(EX\psi)$, we include all states $(\mathbf{s}, \mathbf{x}, \mathbf{Z}_o)$ with the following property: There exists a possible successor set $S_{\mathbf{s}'}$ in which there is a $\mathbf{x}'$, so that for at least one black box output value $\mathbf{Z}'_o$: $(\mathbf{s}', \mathbf{x}', \mathbf{Z}'_o)$ possibly satisfies ψ.

Definition 2.15 We define $Sat_E(EX\psi)$ by

$$\chi_{Sat_E(EX\psi)}(\mathbf{s}, \mathbf{x}, \mathbf{Z}_o) := \exists \mathbf{s}'\Big(\chi_{R_E}(\mathbf{s}, \mathbf{x}, \mathbf{Z}_o, \mathbf{s}') \cdot \exists \mathbf{x}' \exists \mathbf{Z}'_o \big(\chi_{Sat_E(\psi)}|_{\substack{\mathbf{s}\leftarrow\mathbf{s}' \\ \mathbf{x}\leftarrow\mathbf{x}' \\ \mathbf{Z}_o\leftarrow\mathbf{Z}'_o}}\big)(\mathbf{s}', \mathbf{x}', \mathbf{Z}'_o)\Big).$$

Similarly, for $Sat_A(EX\psi)$, we include all states $(\mathbf{s}, \mathbf{x}, \mathbf{Z}_o)$, for which in all possible successor sets $S_{\mathbf{s}'}$ there is a $\mathbf{x}'$, so that for all black box output values $\mathbf{Z}'_o$: $(\mathbf{s}', \mathbf{x}', \mathbf{Z}'_o)$ definitely satisfies ψ, i.e.,

Definition 2.16 We define $Sat_A(EX\psi)$ by

$$\chi_{Sat_A(EX\psi)}(\mathbf{s}, \mathbf{x}, \mathbf{Z}_o) := \forall \mathbf{s}'\Big(\chi_{R_E}(\mathbf{s}, \mathbf{x}, \mathbf{Z}_o, \mathbf{s}') \rightarrow \exists \mathbf{x}' \forall \mathbf{Z}'_o \big(\chi_{Sat_A(\psi)}|_{\substack{\mathbf{s}\leftarrow\mathbf{s}' \\ \mathbf{x}\leftarrow\mathbf{x}' \\ \mathbf{Z}_o\leftarrow\mathbf{Z}'_o}}\big)(\mathbf{s}', \mathbf{x}', \mathbf{Z}'_o)\Big).$$

Lemma 2.8 *Let $\forall \mathbf{Z}_o \chi_{Sat_A(\psi)}$ represent an under-approximation of $Sat_A^{exact}(\psi)$ and let $\exists \mathbf{Z}_o \chi_{Sat_E(\psi)}$ represent an over-approximation of $Sat_E^{exact}(\psi)$. For $Sat_A(EX\psi)$ as defined in Definition 2.16 and for $Sat_E(EX\psi)$ as defined in Definition 2.15, the following holds:*

$$\forall \mathbf{Z}_o \chi_{Sat_A(EX\psi)} \leq \chi_{Sat_A^{exact}(EX\psi)} \text{ and } \chi_{Sat_E^{exact}(EX\psi)} \leq \exists \mathbf{Z}_o \chi_{Sat_E(EX\psi)}.$$

Again, the proof of the lemma follows from the considerations given above the definitions.

The computation of all remaining CTL operators $\neg$, EG and EU is performed as already described before.

Theorem 2.2 *The result of the recursive computation can be evaluated as follows:*

$$\begin{aligned} \big(\forall \mathbf{x} \forall \mathbf{Z}_o (\chi_{Sat_A(\varphi)}|_{\mathbf{s}=\mathbf{s}^0})\big) = 1 &\implies \varphi \text{ is valid} \\ \big(\exists \mathbf{x} \forall \mathbf{Z}_o (\overline{\chi_{Sat_E(\varphi)}}|_{\mathbf{s}=\mathbf{s}^0})\big) = 1 &\implies \varphi \text{ is not realizable.} \end{aligned}$$

Proof The proof follows from Lemmas 2.2, 2.7, and 2.8.

Obviously, including all Z_i-variables into the state space is one extreme case of the method presented in this section which leads to the tightest over- and under-approximations of $Sat_A(\cdot)$ and $Sat_E(\cdot)$. If we include only a part of the Z_i-variables into the state space, then smaller sets of states and transitions have to be considered, which can lead to a less complex model checking run without necessarily losing the accuracy needed for solving the problem.

(5) Exact symbolic model checking for black boxes with bounded memory
Despite the methods described so far being approximate, they are able to disprove realizability and prove validity in many practically relevant cases [24]. As already mentioned in Sect. 2.4, the general decision problem with several black boxes is undecidable [26] however. [24] presents a concept how to provide an *exact* solution to a restricted problem by means of a *conventional symbolic model checker*. Under assumption of an upper bound to the number of the internal states of the black boxes ('bounded memory'), the exact set of black box replacements for which a property is satisfied can be symbolically computed, i.e., an exact answer to both the realizability and the validity question can be given under the bounded memory assumption. The algorithm is based on the extraction of the memory out of the black boxes and (conceptually) on considering all possible choices for the black box instantiations in parallel by means of symbolic methods.

2.5 Conclusion

In this chapter, we have provided a comprehensive study of the verification of incomplete combinational and sequential circuits. We have presented different methods for modeling black boxes: (1) 01X-logic, which is efficient, but pessimistic and over-estimates the set of signals in the circuit which carry an unknown value, (2) quantified Boolean formulas, which constitute an exact formalism for combinational circuits with a single black box, but which are incomplete in other cases, and (3) dependency-quantified Boolean formulas, which are accurate for both combinational and sequential circuits with an arbitrary number of black boxes as long as an upper bound on the amount of memory in the black boxes can be given in advance. We have provided algorithms trading off complexity and precision for (1) the partial equivalence checking (PEC) problem of combinational circuits and (2) the realizability (validity) of invariant properties and general CTL properties for incomplete sequential circuits. For invariant properties we presented algorithms based on SAT, QBF and DQBF solvers, for CTL properties algorithms based on symbolic BDD or AIG representations. If the problems are proven to be realizable by our DQBF based method, our solver HQS is also able to extract corresponding implementations for the black boxes as Skolem functions for the existential variables [31]. An interesting open problem is the question how to compute certificates for *unrealizability* as well for checking with a separate proof checker. Another objective for future work is further increasing the efficiency of the underlying SAT, QBF, and DQBF solvers to improve the scalability of the approach for the approximative as well as the exact methods. Moreover, it will be interesting to look into the generalization of other successful verification methods such as Property Directed Reachability (PDR or IC^3) [5, 11] from complete to incomplete circuits.

References

1. A. Biere, Resolve and expand, in *International Conference on Theory and Applications of Satisfiability Testing (SAT), Vancouver, BC, Canada* (2004)
2. A. Biere, M. Heule, H. van Maaren, T. Walsh (ed.), in *Handbook of Satisfiability*. Frontiers in Artificial Intelligence and Applications, vol. 185 (IOS Press, 2008)
3. R. Bloem, U. Egly, P. Klampfl, R. Könighofer, F. Lonsing, SAT-based methods for circuit synthesis, in *International Conference on Formal Methods in Computer Aided Design (FMCAD), Lausanne, Switzerland* (IEEE, 2014), pp. 31–34
4. R. Bloem, R. Könighofer, M. Seidl, SAT-based synthesis methods for safety specs, in *International Conference on Verification, Model Checking, and Abstract Interpretation (VMCAI)*, ed. By K.L. McMillan, X. Rival. LNCS, vol. 8318 (Springer, San Diego, CA, USA, 2014), pp. 1–20
5. A.R. Bradley, SAT-based model checking without unrolling, in *International Conference on Verification, Model Checking, and Abstract Interpretation (VMCAI)*. LNCS, vol. 6538 (Springer, 2011), pp. 70–87
6. R.E. Bryant, Graph-based algorithms for Boolean function manipulation. IEEE Trans. Comput. Aided Des. **35**(8), 677–691 (1986)
7. R.E. Bryant, Symbolic Boolean manipulation with ordered binary decision diagrams. ACM Comput. Surv. **24**, 293–318 (1992)
8. J.R. Burch, E.M. Clarke, K.L. McMillan, D.L. Dill, L.J. Hwang, Symbolic model checking: 10^{20} states and beyond. Inf. Comput. **98**(2), 142–170 (1992)
9. E.M. Clarke, E.A. Emerson, A.P. Sistla, Automatic verification of finite-state concurrent systems using temporal logic specifications. ACM Trans. Program. Lang. Syst. **8**(2), 244–263 (1986)
10. S.A. Cook, The complexity of theorem-proving procedures, in *Annual ACM Symposium on Theory of Computing (STOC)* (ACM Press, 1971), pp. 151–158
11. N. Eén, A. Mishchenko, R.K. Brayton, Efficient implementation of property directed reachability, in *International Conference on Formal Methods in Computer Aided Design (FMCAD)* (FMCAD Inc., 2011), pp. 125–134
12. A. Fröhlich, G. Kovásznai, A. Biere, H. Veith, iDQ: Instantiation-based DQBF solving, in *International Workshop on Pragmatics of SAT (POS)*, ed. By D.L. Berre. EPiC Series, vol. 27 (EasyChair, Vienna, Austria, 2014), pp. 103–116
13. K. Gitina, S. Reimer, M. Sauer, R. Wimmer, C. Scholl, B. Becker, Equivalence checking for partial implementations revisited, in *Workshop "Methoden und Beschreibungssprachen zur Modellierung und Verifikation von Schaltungen und Systemen" (MBMV)*, ed. By C. Haubelt, D. Timmermann (Universität Rostock, ITMZ, Rostock, Germany, 2013), pp. 61–70
14. K. Gitina, S. Reimer, M. Sauer, R. Wimmer, C. Scholl, B. Becker, Equivalence checking of partial designs using dependency quantified Boolean formulae, in *IEEE International Conference on Computer Design (ICCD), Asheville, NC, USA* (IEEE Computer Society, 2013), pp. 396–403
15. K. Gitina, R. Wimmer, S. Reimer, M. Sauer, C. Scholl, B. Becker, Solving DQBF through quantifier elimination, in *International Conference on Design, Automation and Test in Europe (DATE), Grenoble, France* (IEEE, 2015)
16. E. Giunchiglia, P. Marin, M. Narizzano, sQueezeBF: an effective preprocessor for QBFs based on equivalence reasoning, in *International Conference on Theory and Applications of Satisfiability Testing (SAT)*, ed. By O. Strichman, S. Szeider. LNCS, vol. 6175 (Springer, Edinburgh, UK, 2010), pp. 85–98
17. A. Jain, V. Boppana, R. Mukherjee, J. Jain, M. Fujita, M.S. Hsiao, Testing, verification, and diagnosis in the presence of unknowns, in *IEEE VLSI Test Symposium (VTS)* (IEEE Computer Society, Montreal, Canada, 2000), pp. 263–270
18. S. Jo, A.M. Gharehbaghi, T. Matsumoto, M. Fujita, Debugging processors with advanced features by reprogramming LUTs on FPGA, in *International Conference on Field-Programmable Technology (FPT)* (IEEE, Kyoto, Japan, 2013), pp. 50–57

19. S. Jo, T. Matsumoto, M. Fujita, Sat-based automatic rectification and debugging of combinational circuits with LUT insertions. IPSJ Trans. Syst. LSI Des. Methodol. **7**, 46–55 (2014)
20. K.L. McMillan, *Symbolic Model Checking* (Kluwer Academic Publisher, 1993)
21. A.R. Meyer, L.J. Stockmeyer, Word problems requiring exponential time: preliminary report, in *Annual ACM Symposium on Theory of Computing (STOC)* (ACM Press, 1973), pp. 1–9
22. C. Miller, S. Kupferschmid, M.D.T. Lewis, B. Becker, Encoding techniques, Craig interpolants and bounded model checking for incomplete designs, in *International Conference on Theory and Applications of Satisfiability Testing (SAT)*. LNCS, vol. 6175 (Springer, 2010), pp. 194–208
23. C. Miller, C. Scholl, B. Becker, Proving QBF-hardness in bounded model checking for incomplete designs, in *International Workshop on Microprocessor Test and Verification (MTV)* (IEEE Computer Society, Austin, TX, USA, 2013)
24. T. Nopper, C. Scholl, Symbolic model checking for incomplete designs with flexible modeling of unknowns. IEEE Trans. Comput. **62**(6), 1234–1254 (2013)
25. G. Peterson, J. Reif, S. Azhar, Lower bounds for multiplayer non-cooperative games of incomplete information. Comput. Math. Appl. **41**(7–8), 957–992 (2001)
26. A. Pnueli, R. Rosner, Distributed systems are hard to synthesize, in *IEEE Symposium on Foundations of Computer, Science* (1990), pp. 746–757
27. C. Scholl, B. Becker, Checking equivalence for partial implementations, in *ACM/IEEE Design Automation Conference (DAC)* (ACM Press, Las Vegas, NV, USA, 2001), pp. 238–243
28. A. Smith, A.G. Veneris, M.F. Ali, A. Viglas, Fault diagnosis and logic debugging using boolean satisfiability. IEEE Trans. CAD Integr. Circuits Syst. **24**(10), 1606–1621 (2005)
29. A. Sülflow, G. Fey, R. Drechsler, Using QBF to increase accuracy of SAT-based debugging, in *International Symposium on Circuits and Systems (ISCAS)* (IEEE, Paris, France, 2010), pp. 641–644
30. G.S. Tseitin, On the complexity of derivation in propositional calculus. Stud. Constr. Math. Math. Logic Part **2**, 115–125 (1970)
31. K. Wimmer, R. Wimmer, C. Scholl, B. Becker, Skolem functions for DQBF, in *International Symposium on Automated Technology for Verification and Analysis (ATVA)*, ed. By C. Artho, A. Legay, D. Peled. LNCS, vol. 9938 (Springer, Chiba, Japan, 2016)

Author Biographies

Bernd Becker received the Diploma in Mathematics in 1979, the Doctoral and the Habilitation degree in Computer Science in 1982 and 1988, respectively, all from Saarland University. In 1989, he joined the Institut für Informatik at J.W.Goethe- University Frankfurt as an Associate Professsor for "Complexity Theory and Efficient Algorithms". Since 1995, he is a Full Professor (Chair of Computer Architecture) at the Faculty of Engineering, University of Freiburg. His research activities include design, test and verification methods for embedded systems and nanoelectronic circuitry. Bernd Becker was a Co-Speaker of the DFG Transregional Collaborative Research Center "Automatic Analysis and Verification of Complex Systems (AVACS)" from 2004 to 2015 and is a Director of the Centre for Security and Society, University of Freiburg. He is a fellow of IEEE and Member of Academia Europaea.

Christoph Scholl received the Dipl.-Inform. and the Dr.-Ing. degrees in computer science from University of Saarland, Germany, in 1993 and 1997, respectively. In 2002 he received the venia legendi from University of Freiburg, Germany. In 2002/2003 he was an associate professor for computer engineering at the University of Heidelberg and in 2003 he joined the University of Freiburg as an associate professor in the Department of Computer Science. His research interests include logic synthesis, real-time operating systems, and the verification both of digital circuits and systems and of timed and hybrid systems. In this context a main focus of his work lies on the development of efficient symbolic data structures and algorithms as well as new solver techniques.

Ralf Wimmer received his diploma with distinction in computer science from the Albert-Ludwigs-Universität Freiburg, Germany in 2004. Afterwards, he worked as a Ph.D. student at the Chair of Computer Architecture at the same university, advised by Prof. Dr. Bernd Becker. He obtained his Ph.D. degree with distinction in 2011 for his thesis on symbolic methods for probabilistic verification. Since then, he is continuing his work as a research assistant and leader of the verification group at the Chair of Computer Architecture. His research focus is on symbolic methods and solver technologies, and their application for the verification of digital and stochastic systems.

Chapter 3
Probabilistic Model Checking: Advances and Applications

Marta Kwiatkowska, Gethin Norman and David Parker

3.1 Introduction

Computer systems play an important role in almost all aspects of everyday life, including many examples where safety and reliability are critical, from control systems for autonomous vehicles to embedded software in medical devices such as cardiac pacemakers. There is therefore a demand for rigorous, formal techniques which can verify that these systems function correctly and safely. Often, this requires an analysis of quantitative aspects such as reliability, responsiveness and resource usage. Furthermore, since such devices often operate in unpredictable and unknown environments, it is essential to consider the inherently probabilistic nature of real systems, such as the random timing of events, failures of embedded components and the loss of packets when using wireless communication networks.

Probabilistic model checking is an automated technique for formally verifying quantitative properties of stochastic systems. This involves the construction of a mathematical model that represents the behaviour of a system over time, i.e. the possible states that it can be in, the transitions that can occur between states, and information about the likelihood or timing of these transitions. Properties specifying the required behaviour of these systems are then formally specified in temporal logic and a systematic exploration and analysis of the system model is then performed to ascertain whether the properties are satisfied.

M. Kwiatkowska (✉)
Department of Computer Science, University of Oxford, Oxford, UK
e-mail: marta.kwiatkowska@cs.ox.ac.uk

G. Norman
School of Computing Science, University of Glasgow, Glasgow, UK
e-mail: gethin.norman@glasgow.ac.uk

D. Parker
School of Computer Science, University of Birmingham, Birmingham, UK
e-mail: d.a.parker@cs.bham.ac.uk

R. Drechsler (ed.), *Formal System Verification*, DOI 10.1007/978-3-319-57685-5_3

This approach allows a wide variety of quantitative properties to be specified, regarding, for example, 'the probability of a system failure occurring', 'the probability of a packet being successfully delivered within 5 ms' or 'the expected power consumption of a sensor network during 1 h of operation'. The basic theory and algorithms for probabilistic model checking were first put forward in the 1980s but, since then, substantial progress has been made in the development of theory, algorithms and tools for many different types of probabilistic models and a wide range of property specifications. This has resulted in the successful usage of probabilistic model checking on a huge range of computerised systems, from airbag controllers to cardiac pacemakers, and in a diverse range of applications domains, from computer security to robotics to quantum computing.

This chapter aims to provide both an introduction to the basics of probabilistic model checking and a survey of some of the key advances that have been made in recent years. In both cases, we illustrate the ideas using a variety of toy examples and real-life case studies, and provide pointers to further work and resources. We also make available electronic copies of the files needed to study these examples and case studies using the PRISM model checker [115].

In the first section of the chapter, we give an introduction to probabilistic model checking applied to several different types of models: discrete-time Markov chains, Markov decision processes and stochastic multi-player games. We then move on to cover a section of more advanced topics. This includes: (i) controller synthesis, which can be used to generate correct-by-construction controllers, e.g. for robots or vehicles, along with quantitative guarantees on their behaviour; (ii) modelling and verification techniques designed for large complex systems, including compositional (divide and conquer) approaches and the use of abstraction; (iii) verification techniques for real-time probabilistic models, i.e. those that capture more realistic information about the timing and duration of system events; and (iv) parametric model checking methods, which provide more powerful ways to analyse models whose parameters (e.g. probabilities) may vary or be difficult to quantify accurately. We conclude the chapter with a discussion of the limitations of probabilistic model checking and some of the key current challenges and research directions.

3.2 Probabilistic Model Checking

In this section, we give an overview of the basics of probabilistic model checking. We focus on *discrete-time* models: discrete-time Markov chains (DTMCs), Markov decision processes (MDPs) and stochastic multi-player games (SMGs). These all model the behaviour of a probabilistic system as a sequence of discrete time-steps. We introduce the key definitions and concepts, and illustrate them with some examples. For more in-depth tutorial material on probabilistic model checking, see for example [78] (for DTMCs), [44] (for MDPs) and [104] (for SMGs).

Preliminaries. Before we start, we first introduce some definitions and notation used in the following sections. A (discrete) *probability distribution* over a countable set S is a function $\mu : S{\rightarrow}[0,1]$ such that $\sum_{s \in S} \mu(s) = 1$. For an arbitrary set S, we let $Dist(S)$ be the set of functions $\mu : S{\rightarrow}[0,1]$ such that $\{s \in S \mid \mu(s){>}0\}$ is a countable set and μ restricted to $\{s \in S \mid \mu(s){>}0\}$ is a probability distribution. The *point distribution* at $s \in S$, denoted η_s, is the distribution that assigns probability 1 to s (and 0 to everything else). Given two sets S_1 and S_2 and distributions $\mu_1 \in Dist(S_1)$ and $\mu_2 \in Dist(S_2)$, the *product distribution* $\mu_1{\times}\mu_2 \in Dist(S_1{\times}S_2)$ is given by $\mu_1{\times}\mu_2((s_1,s_2)) = \mu_1(s_1){\cdot}\mu_2(s_2)$. We will also often use the more general notion of a *probability measure*. We omit a complete definition here and instead refer the reader to, for example, [21] for introductory material on this topic.

3.2.1 Discrete-Time Markov Chains

We now give an overview of probabilistic model checking for discrete-time Markov chains, the simplest class of models that we consider in this chapter.

Definition 3.1 (*Discrete-time Markov chain*) A *discrete-time Markov chain* (DTMC) is a tuple $\mathsf{D} = (S, \bar{s}, \mathbf{P}, L)$ where:

- S is a set of states;
- $\bar{s} \in S$ is an initial state;
- $\mathbf{P} : S{\times}S \rightarrow [0,1]$ is a probabilistic transition matrix such that $\sum_{s' \in S} \mathbf{P}(s,s') = 1$ for all $s \in S$;
- $L : S{\rightarrow}2^{AP}$ is a labelling function assigning to each state a set of atomic propositions from a set AP.

The state space S of a DTMC $\mathsf{D} = (S, \bar{s}, \mathbf{P}, L)$ represents the set of all possible configurations of the system being modelled. The system's initial configuration is given by $\bar{s}$ and its subsequent evolution is represented by the probabilistic transition matrix $\mathbf{P}$: for states $s, s' \in S$, the entry $\mathbf{P}(s,s')$ is the probability of making a transition from state s to s'. By definition, for any state of D, the probabilities of all outgoing transitions from that state sum to 1.

A possible execution of D is represented by a *path*, which is a (finite or infinite) sequence of states $\pi = s_0 s_1 s_2 \ldots$ such that $\mathbf{P}(s_i, s_{i+1}){>}0$ for all $i{\geqslant}0$. For a path π, we let $\pi(i)$ denote the $(i{+}1)$th state s_i of the path, and $\pi[i\ldots]$ be the suffix of π starting in state s_i. We also let $|\pi|$ be its length and, if π is finite, $last(\pi)$ be its last state. We let $IPaths_{\mathsf{D}}(s)$ and $FPaths_{\mathsf{D}}(s)$ denote the sets of finite and infinite paths of D starting in state s, respectively, and we write $IPaths_{\mathsf{D}}$ and $FPaths_{\mathsf{D}}$ for the sets of *all* finite and infinite paths, respectively.

To reason quantitatively about the behaviour of DTMC D we must determine the probability that certain paths are executed. To do so, we define, for each state s of D, a probability measure $Pr_{\mathsf{D},s}$ over the set of infinite paths of D starting in s. We present just the basic idea here; for the complete construction, see [76].

For any finite path $\pi = s s_1 s_2 \ldots s_n \in FPaths_{\mathsf{D}}(s)$, the probability of the path occurring is given by $\mathbf{P}(\pi) = \mathbf{P}(s,s_1) \cdot \mathbf{P}(s_1,s_2) \cdots \mathbf{P}(s_{n-1},s_n)$. The *cylinder set* of π, denoted $C(\pi)$, is the set of all infinite paths which have π as a prefix, and the probability assigned to this set of paths is $Pr_{\mathsf{D},s}(C(\pi)) = \mathbf{P}(\pi)$. This can be extended uniquely to define the probability measure $Pr_{\mathsf{D},s}$ over $IPaths_{\mathsf{D}}(s)$.

Using this probability measure, we can quantify the probability that, starting from a state s, the behaviour of D satisfies a particular specification (assuming that the behaviour of interest is represented by a measurable set of paths). For example, we can consider the probability of reaching a particular class of states, or of visiting some set of states infinitely often. Furthermore, given a random variable f over the infinite paths $IPaths_{\mathsf{D}}$ (i.e. a real-valued function $f : IPaths_{\mathsf{D}} \rightarrow \mathbb{R}_{\geqslant 0}$), we can define, using the probability measure $Pr_{\mathsf{D},s}$, the *expected value* of the variable f when starting in s, denoted $\mathbb{E}_{\mathsf{D},s}(f)$. More formally, we have:

$$\mathbb{E}_{\mathsf{D},s}(f) \stackrel{\text{def}}{=} \int_{\pi \in IPaths_{\mathsf{D}}(s)} f(\pi) \mathrm{d}Pr_{\mathsf{D},s} \, .$$

We use random variables to formalise a variety of other quantitative properties of DTMCs. We do so by annotating the model with *rewards* (sometimes, these in fact represent *costs*, but we will consistently refer to these as rewards). Rewards can be used to model, for example, the energy consumption of a device, or the number of packets lost by a communication protocol. Formally, these are defined as follows.

Definition 3.2 (*DTMC reward structure*) A *reward structure* for a DTMC $\mathsf{D} = (S, \bar{s}, \mathbf{P}, L)$ is a tuple $r = (r_S, r_T)$ where $r_S : S \rightarrow \mathbb{R}_{\geqslant 0}$ is a *state reward function* and $r_T : S \times S \rightarrow \mathbb{R}_{\geqslant 0}$ is a *transition reward function*.

State rewards are also called cumulative rewards and transition rewards are sometimes known as instantaneous or impulse rewards. We use random variables to measure, for example, the expected total amount of reward cumulated (over some number of steps, until a set of states is reached, or indefinitely) or the expected value of a reward structure at a particular instant.

Example 1 We now introduce a running example, which we will develop throughout the chapter. It concerns a robot moving through terrain that is divided up into a 3×2 grid, with each grid section represented as a state. Figure 3.1 shows a DTMC model of the robot. In each of the 6 states, the robot selects, at random, a direction to move. Due to the presence of obstacles, certain directions are unavailable in some states. For example, in state s_0, the robot will either remain in its current location (with probability 0.2), move east (with probability 0.35), move south (with probability 0.4) or move south-east (with probability 0.05). We also show labels for the states, taken from the set of atomic propositions $AP = \{\mathsf{hazard}, \mathsf{goal}_1, \mathsf{goal}_2\}$. ■

3.2.1.1 Property Specifications

In order to formally specify properties of interest of a DTMC, we use quantitative extensions of *temporal logic*. For the purposes of this presentation, we introduce a

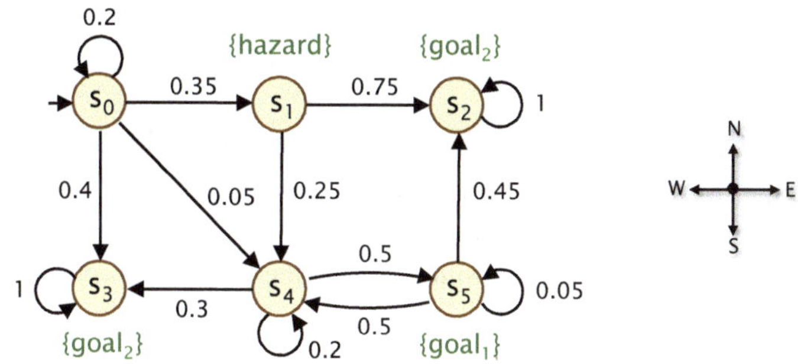

Fig. 3.1 Running example: a DTMC D representing a robot moving about a 3×2 grid

rather general logic that essentially coincides with the property specification language of the PRISM model checker [80]. We refer to it here as *the PRISM logic*. This extends the probabilistic temporal logic PCTL* with operators to specify expected reward properties. PCTL*, in turn, subsumes the logic PCTL (probabilistic computation tree logic) [60] and probabilistic LTL (linear time logic) [95].

Definition 3.3 (*PRISM logic syntax*) The syntax of our logic is given by:

$$\phi ::= \texttt{true} \mid \mathsf{a} \mid \neg\phi \mid \phi \wedge \phi \mid \mathtt{P}_{\bowtie p}[\psi] \mid \mathtt{R}^{r}_{\bowtie q}[\,\rho\,]$$
$$\psi ::= \phi \mid \neg\psi \mid \psi \wedge \psi \mid X\,\psi \mid \psi\ \mathtt{U}^{\leqslant k}\ \psi \mid \psi\ \mathtt{U}\ \psi$$
$$\rho ::= \mathtt{I}^{=k} \mid \mathtt{C}^{\leqslant k} \mid \mathtt{C} \mid \mathtt{F}\,\phi$$

where $\mathsf{a} \in AP$ is an atomic proposition, $\bowtie \in \{<, \leqslant, \geqslant, >\}$, $p \in [0, 1]$, r is a reward structure, $q \in \mathbb{R}_{\geqslant 0}$ and $k \in \mathbb{N}$.

The syntax in Definition 3.3 distinguishes between state formulae (ϕ), path formulae (ψ) and reward formulae (ρ). State formulae are evaluated over the states of a DTMC, while path and reward formulae are both evaluated over paths. A property of a DTMC is specified as a state formula; path and reward formulae appear only as subformulae, within the P and R operators, respectively.

For a state s of a DTMC D, we say that s *satisfies* ψ (or ψ *holds* in s), written D, $s \models \psi$, if ψ evaluates to true in s. If the model D is clear from the context, we simply write $s \models \psi$. In addition to the standard operators of propositional logic, state formulae ϕ can include the probabilistic operator P and reward operator R, which have the following meanings:

- s satisfies $\mathtt{P}_{\bowtie p}[\psi]$ if the probability of taking a path from s satisfying ψ is in the interval specified by $\bowtie p$;
- s satisfies $\mathtt{R}^{r}_{\bowtie q}[\,\rho\,]$ if the expected value of reward operator ρ from state s, using reward structure r, is in the interval specified by $\bowtie q$.

The core temporal operators used to construct path formulae ψ are:

- $X\ \psi$ ('next') – ψ holds in the next state;
- $\psi_1\ \mathtt{U}^{\leqslant k}\ \psi_2$ ('bounded until') – ψ_2 becomes true within k steps, and ψ_1 holds up until that point;
- $\psi_1\ \mathtt{U}\ \psi_2$ ('until') – ψ_2 eventually becomes true, and ψ_1 holds until then.

We often use the equivalences $\mathtt{F}\ \psi \equiv \mathtt{true}\ \mathtt{U}\ \psi$ ('eventually' ψ) and $\mathtt{G}\ \psi \equiv \neg\mathtt{F}\ \neg\psi$ ('always' ψ), as well as the bounded variants $\mathtt{F}^{\leqslant k}\ \psi$ and $\mathtt{G}^{\leqslant k}\ \psi$. When restricting ψ to be an atomic proposition, we get the following common classes of property:

- $\mathtt{F}$ a ('reachability') – eventually a holds;
- $\mathtt{G}$ a ('invariance') – a remains true forever;
- $\mathtt{F}^{\leqslant k}$ a ('step-bounded reachability') – a becomes true within k steps;
- $\mathtt{G}^{\leqslant k}$ a ('step-bounded invariance') – a remains true for k steps.

More generally, path formulae allow temporal operators to be combined. In fact the syntax of path formulae ψ given in Definition 3.3 is that of linear temporal logic (LTL) [95].[1] This logic can express a large class of useful properties, core examples of which include:

- $\mathtt{G}\ \mathtt{F}\ \psi$ ('recurrence') – ψ holds infinitely often;
- $\mathtt{F}\ \mathtt{G}\ \psi$ ('persistence') – eventually ψ always holds;
- $\mathtt{G}\ (\psi_1 \rightarrow X\ \psi_2)$ – whenever ψ_1 holds, ψ_2 holds in the next state;
- $\mathtt{G}\ (\psi_1 \rightarrow \mathtt{F}\ \psi_2)$ – whenever ψ_1 holds, ψ_2 holds at some point in the future.

For reward formulae ρ, we allow four operators:

- $\mathtt{I}^{=k}$ ('instantaneous reward') – the state reward at time step k;
- $\mathtt{C}^{\leq k}$ ('bounded cumulative reward') – the reward accumulated over k steps;
- $\mathtt{C}$ ('total reward') – the total reward accumulated (indefinitely);
- $\mathtt{F}\ \phi$ ('reachability reward') – the reward accumulated up until the first time a state satisfying ϕ is reached.

Numerical queries. It is often of more interest to know the actual probability with which a path formula is satisfied or the expected value of a reward formula, than whether or not the probability or expected value meets a particular bound. To allow such queries, we extend the logic of Definition 3.3 to include *numerical* queries of the form $\mathtt{P}_{=?}[\psi]$ or $\mathtt{R}^{r}_{=?}[\,\rho\,]$, which yield the probability that ψ holds and the expected value of reward operator ρ using reward structure r, respectively.

Example 2 We now return to our running example of a robot navigating a grid (see Example 1 and Fig. 3.1) and illustrate some properties specified in the PRISM logic.

- $\mathtt{P}_{\geqslant 1}[\mathtt{F}\ \mathsf{goal}_2]$ – the probability the robot reaches a goal_2 state is 1.
- $\mathtt{P}_{\geqslant 0.9}[\mathtt{G}\ \neg\mathsf{hazard}]$ – the probability it never visits a hazard state is at least 0.9.

[1]The bounded until operator $\psi_1\ \mathtt{U}^{\leqslant k}\ \psi_2$ is not usually included in the syntax of LTL, but it can be derived from other operators so its inclusion is not problematic.

- $\mathtt{P}_{=?}[\neg\mathsf{hazard}\ \mathtt{U}^{\leqslant k}\ (\mathsf{goal}_1 \vee \mathsf{goal}_2)]$ – what is the probability that the robot reaches a state labelled with either goal_1 or goal_2, while avoiding hazard-labelled states, during the first k steps of operation?
- $\mathtt{R}^{r^1}_{\leqslant 4.5}[\,\mathtt{C}^{\leq k}\,]$ where $r^1 = (r^1_S, r^1_T), r^1_S(s) = 1$ if s is labelled hazard and 0 otherwise and $r_T(s, s') = 0$ for all $s, s' \in S$ – the expected number of times the robot visits a hazard labelled state during the first k steps is at most 4.5.
- $\mathtt{R}^{r^2}_{=?}[\mathtt{F}\ (\mathsf{goal}_1 \vee \mathsf{goal}_2)]$ where $r^2 = (r^2_S, r^2_T), r^2_S(s) = 0$ for all $s \in S$ and $r_T(s, s') = 1$ for all $s, s' \in S$ – what is the expected number of steps required for the robot to reach a state labelled goal_1 or goal_2? ■

The formal semantics of the PRISM logic, for DTMCs, is defined as follows.

Definition 3.4 (*PRISM logic semantics for DTMCs*) For a DTMC $\mathsf{D} = (S, \bar{s}, \mathbf{P}, L)$, reward structure $r = (r_S, r_T)$ for D and state $s \in S$, the satisfaction relation $\models$ is defined as follows:

$$
\begin{array}{rcl}
\mathsf{D}, s \models \mathtt{true} & & \text{always} \\
\mathsf{D}, s \models \mathsf{a} & \Leftrightarrow & \mathsf{a} \in L(s) \\
\mathsf{D}, s \models \neg\phi & \Leftrightarrow & \mathsf{D}, s \not\models \phi \\
\mathsf{D}, s \models \phi_1 \wedge \phi_2 & \Leftrightarrow & \mathsf{D}, s \models \phi_1 \ \wedge\ \mathsf{D}, s \models \phi_2 \\
\mathsf{D}, s \models \mathtt{P}_{\bowtie p}[\psi] & \Leftrightarrow & Pr_{\mathsf{D},s}\{\pi \in IPaths_{\mathsf{D}}(s) \mid \mathsf{D}, \pi \models \psi\} \bowtie p \\
\mathsf{D}, s \models \mathtt{R}^r_{\bowtie q}[\,\rho\,] & \Leftrightarrow & \mathbb{E}_{\mathsf{D},s}(rew^r(\rho)) \bowtie q
\end{array}
$$

where for any path $\pi = s_0 s_1 s_2 \ldots \in IPaths_{\mathsf{D}}$:

$$
\begin{array}{rcl}
\mathsf{D}, \pi \models \phi & \Leftrightarrow & \mathsf{D}, s_0 \models \phi \\
\mathsf{D}, \pi \models \neg\psi & \Leftrightarrow & \mathsf{D}, \pi \not\models \psi \\
\mathsf{D}, \pi \models \psi_1 \wedge \psi_2 & \Leftrightarrow & \mathsf{D}, \pi \models \psi_1 \ \wedge\ \mathsf{D}, \pi \models \psi_2 \\
\mathsf{D}, \pi \models X\ \psi & \Leftrightarrow & \mathsf{D}, \psi[1\ldots] \models \psi \\
\mathsf{D}, \pi \models \psi_1\ \mathtt{U}^{\leqslant k}\ \psi_2 & \Leftrightarrow & \exists i \in \mathbb{N}.\,(\,i \leqslant k \wedge \mathsf{D}, \pi[i\ldots] \models \psi_2 \wedge \forall j{<}i.\,(\mathsf{D}, \pi[j\ldots] \models \psi_1)\,) \\
\mathsf{D}, \pi \models \psi_1\ \mathtt{U}\ \psi_2 & \Leftrightarrow & \exists i \in \mathbb{N}.\,(\,\mathsf{D}, \pi[i\ldots] \models \psi_2 \wedge \forall j{<}i.\,(\mathsf{D}, \pi[j\ldots] \models \psi_1)\,)
\end{array}
$$

$$
\begin{array}{rcl}
rew^r(\mathtt{I}^{=k})(\pi) & = & r_S(s_k) \\
rew^r(\mathtt{C}^{\leq k})(\pi) & = & \sum_{j=0}^{k-1}\big(r_S(s_j) + r_T(s_j, s_{j+1})\big) \\
rew^r(\mathtt{C})(\pi) & = & \sum_{j=0}^{\infty}\big(r_S(s_j) + r_T(s_j, s_{j+1})\big) \\
rew^r(\mathtt{F}\ \phi)(\pi) & = & \begin{cases} \infty & \text{if } \forall j \in \mathbb{N}.\, \mathsf{D}, s_j \not\models \phi \\ \sum_{j=0}^{m_\phi - 1}\big(r_S(s_j) + r_T(s_j, s_{j+1})\big) & \text{otherwise} \end{cases}
\end{array}
$$

and $m_\phi = \min\{j \mid \mathsf{D}, s_j \models \phi\}$.

3.2.1.2 Model Checking

Verifying formulae in this logic against a DTMC requires a combination of graph-based algorithms, automata-based methods using deterministic Rabin automata (DRAs) and solving systems of linear equations. The main components of the model checking procedure are computing the probability that a path formula is satisfied and the expected value of a reward formula. Computing the probability that a path formula is satisfied requires first translating the formula into a DRA, finding the *bottom strongly connected components* on the product of the DTMC (informally, these are the sets of states of a DTMC which once entered are never left) and the constructed automaton and finally solving a linear equation system [15]. Computing the expected value of a reward formula, for unbounded cumulative and reachability reward formulae, also involves graph based analysis (either finding the bottom strongly connected components for unbounded cumulative reward properties or finding the states that reach a target with probability 1 for reachability reward properties) and solving a system of linear equations [78]. For the remaining reward formulae, computation of expected values involves iteratively solving a set of linear equations.

The overall complexity of model checking is doubly exponential in the formula and polynomial in the size of the DTMC, but can be reduced to a single exponential. For scalability reasons, when implementing model checking of DTMCs, iterative numerical methods such as Jacobi and Gauss–Seidel, as opposed to direct methods such as Gaussian elimination, are often employed when solving systems of linear equations.

3.2.1.3 Case Study: NAND Multiplexing

We now describe a case study in which the system is modelled as a DTMC. This is taken from [91] and concerns the analysis of defect-tolerant systems used in computer-aided design. The system under study uses *multiplexing*, a technique introduced by von Neumann [109] which enables reliable computations when using unreliable devices. The approach was developed due to the unreliability of the valves (also known as vacuum tubes) that were used in computers, and these techniques are becoming relevant again for systems developed using nanotechnology where, due to their small-scale, components are again unreliable.

Multiplexing involves replacing a single processing unit by a multiplexing unit which has N copies of the inputs and outputs of the original processing unit. In the multiplexing unit, the N inputs are processed in parallel, giving N outputs. If the inputs and devices are reliable, then each of the N outputs would equal the output of the single processing unit. However, if there are errors in the inputs or the processing is unreliable, then there will also be errors in the outputs. To give a value to the output of the multiplexing unit, we define a critical level $\Delta \in [0, 0.5)$ and, if at least $(1-\Delta)\cdot N$ of the outputs take a certain value (i.e. either `true` or `false`), this is taken as the output value. If this criteria is not met by either `true` or `false`, the output value of the multiplexing unit is unknown and an error occurs.

The design of a multiplexing unit comprises an *executive stage*, which carries out the basic function of the unit to be replaced, and M *restorative stages*, which reduce the degradation of the output from the executive stage caused by errors in the inputs and unreliable processing. For the case of NAND multiplexing, the focus of this case study, a design with a single restorative stage is shown in Fig. 3.2.

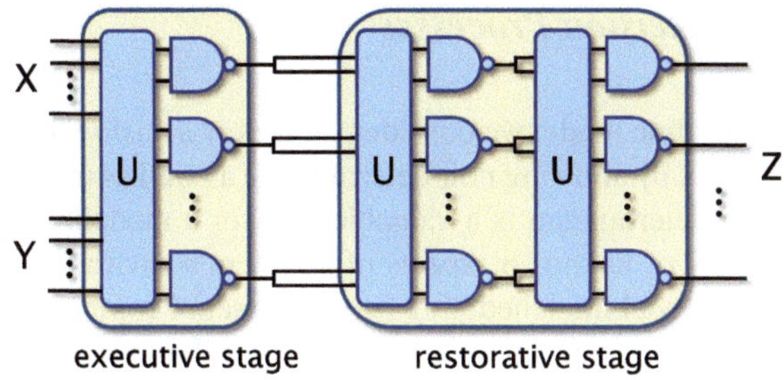

Fig. 3.2 An example of a NAND multiplexing unit with one restorative stage ($M = 1$)

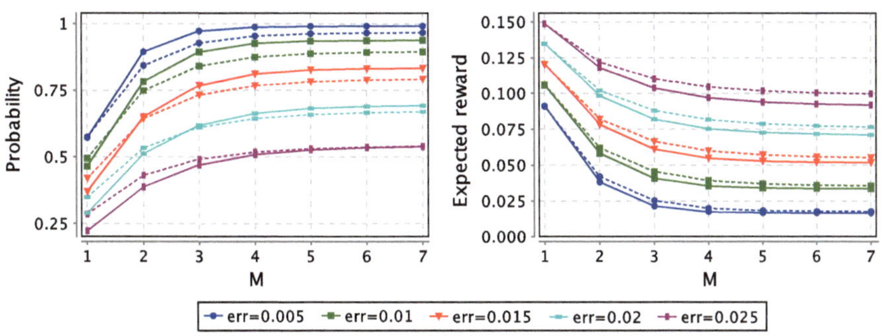

Fig. 3.3 Probabilistic model checking results for the NAND case study

Figure 3.3 presents results obtained with the probabilistic model checker PRISM [80, 114] when analysing: (i) the probability of errors being less than 10%; and (ii) the expected percentage of incorrect outputs of the system. The values are plotted as the number of restorative units (M) and the probability that a NAND gate is unreliable (err) vary. The first property can be expressed as the numerical query $\mathtt{P}_{=?}[\mathtt{F}$ below], where below is an atomic proposition labelling states of the DTMC where the computation has finished and the number of errors is below 10%. The second property can be expressed as the query $\mathtt{R}^{r}_{=?}[\ \mathtt{F}$ done], where done labels states of the DTMC where the computation has completed, and the reward structure r labels the transitions entering this state with a reward equal to the percentage of incorrect outputs. When studying this model with PRISM [91], an error was found in the analytical analysis of [57]. The dashed lines in Fig. 3.3 show the results obtained in this case and demonstrate that this error can cause both under- and over-approximations of the reliability of a NAND multiplexing unit.

3.2.2 Markov Decision Processes

The second discrete-time model we consider is *Markov decision processes* (MDPs). These extend DTMCs by allowing non-deterministic as well as (discrete) probabilistic behaviour. Non-determinism is a valuable tool for a modeller and can be used to represent a variety of unknown aspects of a system's environment or execution. For example, it can model the scheduling between a set of components running concurrently, the instructions and inputs provided to a robot to control its execution, or the unknown behaviour of an adversary trying to attack a security system. More generally, non-determinism can also be used to abstract parts of a system that are unknown, under-specified or unimportant.

Definition 3.5 (*Markov decision process*) A *Markov decision process* (MDP) is a tuple $\mathsf{M} = (S, \bar{s}, A, \delta, L)$ where:

- S is a finite set of states;
- $\bar{s} \in S$ is an initial state;
- A is a finite set of *actions*;
- $\delta : S \times A \rightarrow Dist(S)$ is a (partial) probabilistic transition function, mapping state-action pairs to probability distributions over S;
- $L : S \rightarrow 2^{AP}$ is a state labelling function.

In a state s of an MDP $\mathsf{M} = (S, \bar{s}, A, \delta, L)$, there is first a non-deterministic choice between a set of actions that are *available* in the state. This set, denoted $A(s)$, includes the actions for which a probabilistic transition is defined: $A(s) = \{a \mid \delta(s, a) \text{ is defined}\}$. We assume that the set $A(s)$ is non-empty for all states $s \in S$. Once an available action $a \in A(s)$ has been chosen in s, the action is performed and the successor state s' is chosen probabilistically, where the probability of moving to state s' is $\delta(s, a)(s')$.

Like for DTMCs, a path is a sequence of states corrected by transitions, but now also incorporates the action choice made. A (finite or infinite) path of M is of the form $\pi = s_0 \xrightarrow{a_0} s_1 \xrightarrow{a_1} \cdots$, where $a_i \in A(s_i)$ and $\delta(s_i, a_i)(s_{i+1}) > 0$ for all $i \geqslant 0$. The sets of all finite and infinite paths from state s of M are denoted $FPaths_{\mathsf{M}}(s)$ and $IPaths_{\mathsf{M}}(s)$, respectively, and the sets of all such paths are $FPaths_{\mathsf{M}}$ and $IPaths_{\mathsf{M}}$.

As for DTMCs we can define a reward structure over an MDP. State rewards remain unchanged, however for MDPs, instead of rewards beginning associated with individual transitions, rewards are associated with performing actions in states.

Definition 3.6 (*MDP reward structure*) A *reward structure* for an MDP $\mathsf{M} = (S, \bar{s}, A, \delta, L)$ is a tuple $r = (r_S, r_A)$ where $r_S : S \rightarrow \mathbb{R}_{\geqslant 0}$ is a *state reward function* and $r_A : S \times A \rightarrow \mathbb{R}_{\geqslant 0}$ is an *action reward function*.

Example 3 We now return to our running example of a robot moving through terrain that is divided up into a 3×2 grid (see Example 1 and Fig. 3.1). We extend our earlier DTMC model so that, instead of the robot choosing a direction to move at random, the choice is modelled using non-determinism in an MDP. The model is shown in

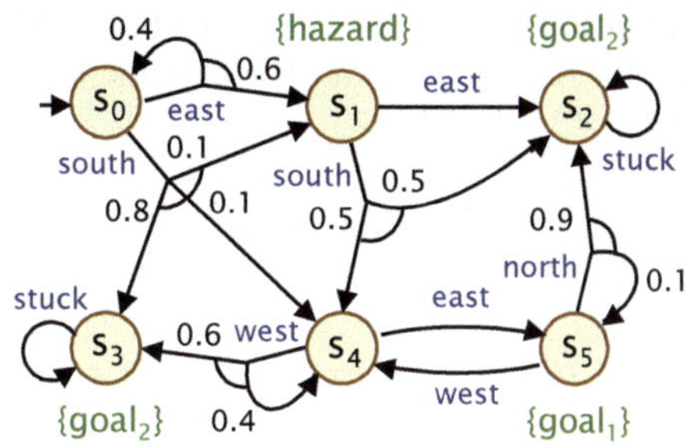

Fig. 3.4 Running example: an MDP M representing a robot moving about a 3×2 grid

Fig. 3.4. The probabilistic transition function is drawn as grouped, labelled arrows and, when the probability is 1, it is omitted. In each state, one or more actions from the set $A = \{north, east, south, west, stuck\}$ are available, which move the robot between grid sections. As for the DTMC model, due to the presence of obstacles, certain directions are unavailable and in this case the obstacles can also cause the robot to probabilistically move to an alternative grid section. We use the action *stuck* to indicate that the robot cannot move in any direction in the states s_2 and s_3. ■

To reason about the behaviour of an MDP, we need the notion of *strategies* (also called policies, adversaries and schedulers in other contexts). A strategy resolves the non-determinism in the model, that is, the choices of which action to perform in each state. This choice can depend on the history of the MDP's execution and can be made either deterministically or randomly.

Definition 3.7 (*MDP strategy*) A *strategy* of an MDP $\mathsf{M} = (S, \bar{s}, A, \delta, L)$ is a function $\sigma : FPaths_{\mathsf{M}} \to Dist(A)$ such that $\sigma(\pi)(a) > 0$ only if $a \in A(last(\pi))$. The set of all strategies of M is denoted Σ_{M}.

We classify strategies in terms of both their use of *randomisation* and of *memory*.

- **Randomisation**: we say that strategy σ is *deterministic* (or *pure*) if $\sigma(\pi)$ is a point distribution for all finite paths π, and *randomised* otherwise.
- **Memory**: a strategy σ is *memoryless* if $\sigma(\pi)$ depends only on $last(\pi)$ for all finite paths π, and *finite-memory* if there are finitely many *modes* such that, for any π, $\sigma(\pi)$ depends only on $last(\pi)$ and the current mode, which is updated each time an action is performed; otherwise, it is *infinite-memory*.

Under a strategy $\sigma \in \Sigma_{\mathsf{M}}$ of MDP M, all non-determinism of M is resolved, and hence the behaviour is fully probabilistic. We can represent this using an (infinite) *induced discrete-time Markov chain*, whose states are the finite paths of M. For a given state s of M, we can then use this DTMC (see Sect. 3.2.1) to construct a probability measure $Pr^{\sigma}_{\mathsf{M},s}$ over the infinite paths $IPaths_{\mathsf{M}}(s)$, capturing the behaviour of M when starting from state s under strategy σ. Furthermore, for a random variable $f : IPaths_{\mathsf{M}} \to \mathbb{R}_{\geqslant 0}$, we can define the expected value $\mathbb{E}^{\sigma}_{\mathsf{M},s}(f)$ of f when starting from state s under strategy σ. Formally, the induced DTMC can be defined as follows.

Definition 3.8 (*Induced DTMC*) For an MDP $\mathsf{M} = (S, \bar{s}, A, \delta, L)$ and strategy $\sigma \in \Sigma_{\mathsf{M}}$ for M, the *induced DTMC* is the DTMC $\mathsf{M}_\sigma = (FPaths_{\mathsf{M}}, \bar{s}, \mathbf{P}, L')$ where, for any $\pi, \pi' \in FPaths_{\mathsf{M}}$:

$$\mathbf{P}(\pi, \pi') = \begin{cases} \sigma(\pi)(a) \cdot \delta(last(\pi), a)(s) & \text{if } \pi' = \pi \xrightarrow{a} s \text{ for some } a \in A \text{ and } s \in S; \\ 0 & \text{otherwise;} \end{cases}$$

and $L'(\pi) = L(last(\pi))$ for all $\pi \in FPaths_{\mathsf{M}}$. Furthermore, a reward structure $r = (r_S, r_A)$ over M induces the reward structure $r^\sigma = (r^\sigma_S, r^\sigma_T)$ over M_σ where for any $\pi, \pi' \in FPaths_{\mathsf{M}}$:

$$\begin{aligned} r^\sigma_S(\pi) &= r_S(last(\pi)) \\ r^\sigma_T(\pi, \pi') &= \begin{cases} r_A(last(\pi), a) & \text{if } \pi' = \pi \xrightarrow{a} s \text{ for some } a \in A \text{ and } s \in S; \\ 0 & \text{otherwise.} \end{cases} \end{aligned}$$

An induced DTMC has an infinite number of states. However, in the case of finite-memory strategies (and hence also the subclass of memoryless strategies), we can construct a finite-state *quotient DTMC* [44].

To specify properties of MDPs, we again use the PRISM logic defined for DTMCs in the previous section. The syntax (see Definition 3.3) is identical, and the semantics (see Definition 3.4) is very similar, the key difference being that, for the $\mathtt{P}_{\bowtie p}[\psi]$ and $\mathtt{R}^r_{\bowtie q}[\,\rho\,]$ operators, we quantify over all possible strategies of the MDP. The treatment of reward operators is also adapted slightly to consider action, as opposed to transition, reward functions.

Definition 3.9 (*PRISM logic semantics for MDPs*) For an MDP $\mathsf{M} = (S, \bar{s}, A, \delta, L)$ and reward structure $r = (r_S, r_A)$ for M, the satisfaction relation $\models$ is defined as for DTMCs in Definition 3.4, except that, for any state $s \in S$:

$$\begin{aligned} \mathsf{M}, s \models \mathtt{P}_{\bowtie p}[\psi] &\Leftrightarrow Pr^\sigma_{\mathsf{M},s}\{\pi \in IPaths_{\mathsf{M}}(s) \mid \mathsf{M}, \pi \models \psi\} \bowtie p \ \text{ for all } \sigma \in \Sigma_{\mathsf{M}} \\ \mathsf{M}, s \models \mathtt{R}^r_{\bowtie q}[\,\rho\,] &\Leftrightarrow \mathbb{E}^\sigma_{\mathsf{M},s}(rew^r(\rho)) \bowtie q \ \text{ for all } \sigma \in \Sigma_{\mathsf{M}} \end{aligned}$$

and, for any path $\pi = s_0 \xrightarrow{a_0} s_1 \xrightarrow{a_1} \cdots \in IPaths_{\mathsf{M}}$:

$$\begin{aligned} rew^r(\mathtt{I}^{=k})(\pi) &= r_S(s_k) \\ rew^r(\mathtt{C}^{\leq k})(\pi) &= \textstyle\sum_{j=0}^{k-1} \big(r_S(s_j) + r_A(s_j, a_j)\big) \\ rew^r(\mathtt{C})(\pi) &= \textstyle\sum_{j=0}^{\infty} \big(r_S(s_j) + r_A(s_j, a_j)\big) \\ rew^r(\mathtt{F}\ \phi)(\pi) &= \begin{cases} \infty & \text{if } \forall j \in \mathbb{N}.\, \mathsf{M}, s_j \not\models \phi \\ \sum_{j=0}^{m_\phi - 1} \big(r_S(s_j) + r_A(s_j, a_j)\big) & \text{otherwise} \end{cases} \end{aligned}$$

where $m_\phi = \min\{j \mid \mathsf{M}, s_j \models \phi\}$.

The main components of the model checking procedure for this logic against an MDP are computing the optimal probabilities that a path formula is satisfied and the optimal expected values for a reward formula. More precisely we are concerned with the following optimal values for an MDP M and state s:

$$Pr^{\min}_{\mathsf{M},s}(\psi) \stackrel{\text{def}}{=} \inf_{\sigma\in\Sigma_{\mathsf{M}}} Pr^{\sigma}_{\mathsf{M},s}\{\pi \in IPaths_{\mathsf{M}}(s) \mid \mathsf{M}, \pi \models \psi\} \tag{3.1}$$

$$Pr^{\max}_{\mathsf{M},s}(\psi) \stackrel{\text{def}}{=} \sup_{\sigma\in\Sigma_{\mathsf{M}}} Pr^{\sigma}_{\mathsf{M},s}\{\pi \in IPaths_{\mathsf{M}}(s) \mid \mathsf{M}, \pi \models \psi\} \tag{3.2}$$

$$\mathbb{E}^{\min}_{\mathsf{M},s}(r,\rho) \stackrel{\text{def}}{=} \inf_{\sigma\in\Sigma_{\mathsf{M}}} \mathbb{E}^{\sigma}_{\mathsf{M},s}(rew^{r}(\rho)) \tag{3.3}$$

$$\mathbb{E}^{\max}_{\mathsf{M},s}(r,\rho) \stackrel{\text{def}}{=} \sup_{\sigma\in\Sigma_{\mathsf{M}}} \mathbb{E}^{\sigma}_{\mathsf{M},s}(rew^{r}(\rho)) \tag{3.4}$$

where ψ is a path formula, r is a reward structure of M and ρ is a reward formula.

For example, verifying the property $\phi = \mathtt{P}_{<p}[\psi]$ in state s of M can be achieved by computing the optimal probability $Pr^{\max}_{\mathsf{M},s}(\psi)$ since the state s satisfies ϕ if and only if $Pr^{\max}_{\mathsf{M},s}(\psi)<p$. Similarly to DTMCs, rather than fixing a specific bound, we can query the (optimal) values directly. In the case of MDPs, the syntax of the PRISM logic is extended to include numerical queries of the form $\mathtt{P}_{\min=?}[\psi]$, $\mathtt{P}_{\max=?}[\psi]$, $\mathtt{R}^{r}_{\min=?}[\,\rho\,]$ and $\mathtt{R}^{r}_{\max=?}[\,\rho\,]$.

Model checking for an MDP reduces to building DRAs, performing graph analysis and numerical computation. As for DTMC model checking, DRAs are built to represent path formulae. The graph analysis involves identifying states of the MDP for which the probability is 0 or 1 and finding *maximal end components* of the MDP (or of the product of the MDP and a DRA). Informally, end components of an MDP are sets of states for which it possible (i.e. assuming certain non-deterministic choices are made) to remain in forever once entered.

The numerical computation can be achieved using various methods including solving a linear programming problem; policy iteration (which builds a sequence of strategies until an optimal one is reached); and value iteration, which computes increasingly precise approximations to the optimal probability or expected value. The overall complexity for model checking is doubly exponential in the formula and polynomial in the size of the MDP.

Further details on the techniques needed to analyse MDPs can be found in, for example, [15, 36, 44] and in standard texts on MDPs [20, 64, 97].

3.2.3 Stochastic Multi-player Games

The final model we consider in this introductory section is *stochastic multi-player games* (SMGs). These extend MDPs by allowing different *players* to resolve the non-determinism (MDPs can thus be considered as 1-player stochastic games). SMGs allow us to reason about the strategic decisions of several agents either competing or collaborating to achieve some objective. We restrict our attention to *turn-based* stochastic games, in which a single player is responsible for the non-deterministic choices available in each state. We have the following formal definition.

Definition 3.10 (*Stochastic multi-player game*) A (turn-based) *stochastic multi-player game* (SMG) is a tuple $\mathsf{G} = (\Pi, S, (S_i)_{i \in \Pi}, \bar{s}, A, \delta, L)$, where:

- $(S, \bar{s}, A, \delta, L)$ represents an MDP (see Definition 3.5);
- Π is a finite set of *players*;
- $(S_i)_{i \in \Pi}$ is a partition of S.

In a state s of an SMG G, the evolution is similar to an MDP: first an available action is non-deterministically chosen and then the successor state is chosen according to the distribution $\delta(s, a)$. The difference is that the non-deterministic choice is resolved by the *player* that controls the state s, that is, the player $i \in \Pi$ for which $s \in S_i$. As for MDPs, we can define the set of finite and infinite paths $FPaths_{\mathsf{G}}$ ($FPaths_{\mathsf{G}}(s)$) and $IPaths_{\mathsf{G}}$ ($IPaths_{\mathsf{G}}(s)$) of G. Furthermore, we can define reward structures for SMGs in the same way as for MDPs (see Definition 3.6).

To resolve the non-determinism in an SMG, we again use strategies, however we now define a separate strategy for each player of the game.

Definition 3.11 (*SMG strategy*) For an SMG $\mathsf{G} = (\Pi, S, (S_i)_{i \in \Pi}, \bar{s}, A, \delta, L)$, a strategy σ_i for player i of G is a function $\sigma_i : \{\pi \mid \pi \in FPaths_{\mathsf{G}} \wedge last(\pi) \in S_i\} \rightarrow Dist(A)$ such that, if $\sigma_i(\pi)(a) > 0$, then $a \in A(last(\pi))$. The set of all strategies for player $i \in \Pi$ in SMG G is denoted by Σ^i_{G}.

For an SMG $\mathsf{G} = (\Pi, S, (S_i)_{i \in \Pi}, \bar{s}, A, \delta, L)$ and strategies $\sigma_1, \ldots, \sigma_k$ for multiple players $1, \ldots, k$, we can combine them into a single strategy $\sigma = \sigma_1, \ldots, \sigma_k$ which controls the non-determinism when the game is in the states $S_1 \cup \cdots \cup S_k$. If a combined strategy σ is constructed from all the players Π of G (sometimes called a *strategy profile*), then the non-determinism is resolved in all the states of the game and, as for MDPs, we can construct probability measures $Pr^{\sigma}_{\mathsf{G},s}$ over the infinite paths of G.

To specify properties of SMGs, we consider an extension of the PRISM logic used earlier for DTMCs and MDPs, adding the *coalition* operator $\langle\langle C \rangle\rangle$ from alternating temporal logic (ATL) [6]. The result is (essentially) the logic RPATL* proposed in [28].[2]

Definition 3.12 (*RPATL* syntax*) The syntax of RPATL* is given by:

$$\phi ::= \texttt{true} \mid \mathsf{a} \mid \neg\phi \mid \phi \wedge \phi \mid \langle\langle C \rangle\rangle \; \mathtt{P}_{\bowtie p}[\psi] \mid \langle\langle C \rangle\rangle \; \mathtt{R}^r_{\bowtie q}[\,\rho\,]$$

where path formulae ψ and reward formulae ρ are defined in identical fashion to the PRISM logic in Definition 3.3, $C \subseteq \Pi$ is a coalition of players, $\mathsf{a} \in AP$, $\bowtie \in \{<, \leqslant, \geqslant, >\}$, $p \in [0, 1]$, r is a reward structure and $q \in \mathbb{R}_{\geqslant 0}$.

Intuitively, the formulae $\langle\langle C \rangle\rangle \; \mathtt{P}_{\bowtie p}[\psi]$ and $\langle\langle C \rangle\rangle \; \mathtt{R}^r_{\bowtie q}[\,\rho\,]$ mean that it is possible for the players in C to collectively ensure that $\mathtt{P}_{\bowtie p}[\psi]$ or $\mathtt{R}^r_{\bowtie q}[\,\rho\,]$, respectively, is satisfied, no matter what the other players of the game decide to do. We can also

[2] Strictly speaking, the definition of reward operators differs in [28].

adapt these to numerical queries, writing for example $\langle\langle C\rangle\rangle\ \mathtt{P}_{\max=?}[\psi]$ to represent the maximum probability of ψ that the players in C can ensure, regardless of the choices of the other players of the game.

In order to formalise the semantics of RPATL*, we define *coalition games.*

Definition 3.13 (*Coalition game*) Given an SMG $\mathsf{G} = (\Pi, S, (S_i)_{i\in\Pi}, \bar{s}, A, \delta, L)$ and coalition of players $C \subseteq \Pi$, the *coalition game* of G induced by C is the stochastic two-player game $\mathsf{G}_C = (\{1,2\}, S, (S'_1, S'_2), \bar{s}, A, \delta, L)$ where $S'_1 = \cup_{i\in C} S_i$ and $S'_2 = \cup_{i\in\Pi\setminus C} S_i$.

Definition 3.14 (*RPATL* semantics*) For an SMG $\mathsf{G} = (\Pi, S, (S_i)_{i\in\Pi}, \bar{s}, A, \delta, L)$ and reward structure $r = (r_S, r_A)$ for G, the satisfaction relation $\models$ is defined as in Definition 3.9 except that, for any state $s \in S$:

$$\begin{array}{rcl} \mathsf{G}, s \models \langle\langle C\rangle\rangle\ \mathtt{P}_{\bowtie p}[\psi] & \Leftrightarrow & \text{there exists} \sigma_1 \in \Sigma^1_{\mathsf{G}_C} \text{ such that, for any } \sigma_2 \in \Sigma^2_{\mathsf{G}_C}, \\ & & \text{we have } Pr^{\sigma_1,\sigma_2}_{\mathsf{G}_C,s}\{\pi \in IPaths_{\mathsf{G}_C}(s) \mid \mathsf{G}, \pi \models \psi\} \bowtie p \\ \mathsf{G}, s \models \langle\langle C\rangle\rangle\ \mathtt{R}^r_{\bowtie q}[\,\rho\,] & \Leftrightarrow & \text{there exists} \sigma_1 \in \Sigma^1_{\mathsf{G}_C} \text{such that, for any} \sigma_2 \in \Sigma^2_{\mathsf{G}_C}, \\ & & \text{we have } \mathbb{E}^{\sigma_1,\sigma_2}_{\mathsf{G}_C,s}(rew^r(\rho)) \bowtie q \end{array}$$

where $\mathsf{G}_C = (S, (S'_1, S'_2), \bar{s}, A, \delta, L)$ is the coalition game of G induced by C.

As can be seen in Definition 3.14, model checking of RPATL* reduces to the analysis of stochastic two-player games. The exact complexity of analysing such games is a long-standing open problem [31], but key properties such as reachability probabilities and expected cumulated rewards can be efficiently approximated using methods such as value iteration [32]. The overall model checking problem can be performed in a similar manner to the algorithms described for model checking MDPs described in Sect. 3.2.2. For further details, see [28]. SMG model checking has been applied to case studies across a number of application domains, including autonomous transport, security protocols and energy management systems. See, for example, [28, 111, 116] for details.

3.2.4 Tool Support

There are several software tools available for probabilistic model checking. One of the most widely used of these is PRISM [80], which incorporates the majority of the techniques covered in this chapter. In particular, it supports model checking of DTMCs and MDPs, as described above, as well as probabilistic automata, continuous-time Markov chains and probabilistic timed automata, which are discussed in later sections. PRISM-games [86] is an extension of PRISM for the verification of SMGs. Another widely used tool is MRMC [73], which can be used to analyse Markov chains and Markov reward models, and also has support for continuous-time MDPs (a model combining non-deterministic, probabilistic and real-time features, see Sect. 3.5). Other general purpose probabilistic model checking tools include the

Modest Toolset [61], iscasMc [55] and PAT [103]. More specialised tools, focusing on techniques such as parametric model checking or abstraction refinement, are mentioned in the corresponding sections of this chapter. A more extensive list of available tools is maintained at [117].

3.3 Controller Synthesis

In this section, we describe a technique that is closely related to probabilistic model checking: *controller synthesis*. For probabilistic models that include non-determinism, such as MDPs and SMGs, there are two, dual ways that we can reason about them. First, as done in the earlier sections of this chapter, we can *verify* that the model satisfies some formally specified property *for all* possible resolutions of non-determinism. Secondly, we can *synthesise* a controller (i.e. a means of resolving the non-determinism) under which a formally specified property is guaranteed to hold.

In this section, we describe controller synthesis techniques applied to MDPs. For SMGs, model checking of the logic RPATL*, discussed earlier, provides a good basis for controller synthesis in the context of multiple agents. Later, in Sect. 3.5, we will illustrate controller synthesis for real-time probabilistic systems using probabilistic timed automata.

3.3.1 Controller Synthesis for MDPs

To apply controller synthesis to a system modelled as an MDP, we use *strategy synthesis*, which generates a strategy under which a particular formally-specified property is guaranteed to be true. We focus on a subset of the PRISM logic from Definition 3.3 comprising a single instance of a $\mathtt{P}_{\bowtie p}[\cdot]$ or $\mathtt{R}^r_{\bowtie q}[\,\cdot\,]$ operator. In particular, further instances of these operators are not allowed to be nested within path formulae or reward formulae (these cases are known to be more challenging [13, 23]).

A formal definition of strategy synthesis is given below. For this, we use a slightly different form of the satisfaction relation $\models$, where we write $\mathsf{M}, \sigma, s \models \phi$ to state that property ϕ is satisfied by MDP M under the strategy σ (which is essentially the same as the satisfaction of ϕ under the induced DTMC M_σ).

Definition 3.15 (*Strategy synthesis*) The *strategy synthesis* problem is: given an MDP M with initial state $\bar{s}$ and a formula ϕ of the form $\mathtt{P}_{\bowtie p}[\psi]$ or $\mathtt{R}^r_{\bowtie q}[\,\rho\,]$ (see Definition 3.3), find, if it exists, a strategy $\sigma^* \in \Sigma_{\mathsf{M}}$ such that $\mathsf{M}, \sigma^*, \bar{s} \models \phi$.

Like for probabilistic model checking of MDPs, as discussed in Sect. 3.2.2, the problem of strategy synthesis for a $\mathtt{P}_{\bowtie p}[\psi]$ or $\mathtt{R}^r_{\bowtie q}[\,\rho\,]$ operator can be solved by computing an *optimal value* (i.e. minimum or maximum value) for ψ or ρ. For example, when attempting to synthesise a strategy for $\phi = \mathtt{P}_{\leqslant p}[\psi]$, we can compute $Pr^{\min}_{\mathsf{M},\bar{s}}(\psi)$. Then, there exists a strategy σ^* satisfying ϕ if and only if $Pr^{\min}_{\mathsf{M},\bar{s}}(\psi) \leqslant p$, in which case we can take σ^* to be a corresponding *optimal strategy*, i.e. one that achieves the optimal value $Pr^{\min}_{\mathsf{M},\bar{s}}(\psi)$. So, in general, rather than fixing a specific bound p, we can just use a numerical query such as $\mathtt{P}_{\min=?}[\psi]$ to specify a strategy synthesis problem, and directly compute an optimal value and strategy for it.

We already sketched the techniques required to compute optimal values for such properties of MDPs in Sect. 3.2.2. In the sections below, we recap the required computations, additionally discussing which classes of strategies need to be considered for optimality (i.e. the smallest class of strategies guaranteed to contain an optimal one) and the methods required to generate them.

3.3.1.1 Probabilistic Reachability

For probabilistic reachability queries $\mathtt{P}_{\min=?}[\mathtt{F}\ \mathsf{a}]$ or $\mathtt{P}_{\max=?}[\mathtt{F}\ \mathsf{a}]$, *memoryless deterministic* strategies achieve optimal values, and so this class of strategy suffices for strategy synthesis. Determining optimal probability values requires an analysis of the underlying graph structure of the MDP, followed by a numerical computation phase using, for example, linear programming, policy iteration or value iteration.

The construction of an optimal strategy σ^* then depends on the method used in the numerical computation phase. Policy iteration is the most direct as an optimal strategy is constructed as part of the algorithm. For the remaining methods, the optimal strategy corresponds to selecting locally optimal actions in each state, although maximum probabilities require care.

In the case of a bounded reachability query $\mathtt{P}_{\min=?}[\mathtt{F}^{\leqslant k}\ \mathsf{a}]$ or $\mathtt{P}_{\max=?}[\mathtt{F}^{\leqslant k}\ \mathsf{a}]$, memoryless strategies do not achieve optimal values and instead we need to consider the larger class of *finite-memory deterministic* strategies. Strategy synthesis and the computation of optimal reachability probabilities for step-bounded reachability corresponds to working backwards through the MDP and determining, at each step, the actions that yields optimal probabilities in each state.

Example 4 We now return to the MDP M from Fig. 3.4 and synthesise a strategy for the numerical probabilistic reachability query $\mathtt{P}_{\max=?}[\mathtt{F}\ \mathsf{goal}_1]$. Therefore, we first compute the optimal value $Pr^{\max}_{\mathsf{M},s_0}(\mathtt{F}\ \mathsf{goal}_1)$, which we find equals 0.5. Synthesising an optimal strategy, we find the memoryless deterministic strategy (see Fig. 3.5) that selects *east* in s_0, *south* in s_1 and *east* in s_4 (there is no choice needed in s_2 or s_3, and the choice in s_5 is not relevant as the target goal_1 has been reached).

Next, consider the bounded probabilistic reachability query $\mathtt{P}_{\max=?}[\mathtt{F}^{\leqslant k}\ \mathsf{goal}_2]$. We find that the maximum probability equals 0.8, 0.96 and 0.99 for $k = 1, 2$ and 3, respectively. In the case where $k = 3$, the synthesised strategy is deterministic and finite-memory. In particular the strategy, when arriving in state s_4, after 1 step, selects *east* (since goal_2 is reached with probability 0.9). On the other hand, arriving in state s_4 after 2 steps, the strategy selects *west* (since otherwise goal_2 cannot be reached within $k-2 = 1$ steps). ■

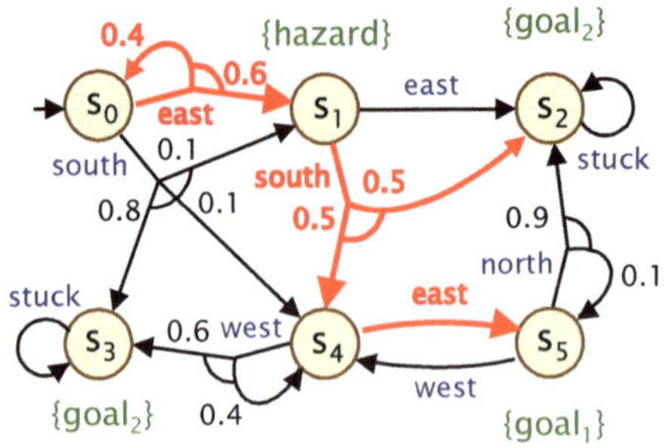

Fig. 3.5 Running example: an MDP M representing a robot moving about a 3×2 grid

3.3.1.2 Reward Properties

Strategy synthesis for a numerical reward query $\mathtt{R}^{r}_{\min=?}[\,\rho\,]$ or $\mathtt{R}^{r}_{\max=?}[\,\rho\,]$ is similar to probabilistic reachability queries. In the case of reachability rewards, i.e. when ρ is of the form $\mathtt{F}\ \mathsf{a}$, as for unbounded probabilistic reachability, first there is a precomputation phase (identifying the states for which the expected reward is infinite), and then a numerical computation phase using methods such as value iteration, policy iteration or linear programming. As for unbounded probabilistic reachability, it is sufficient to consider the class of memoryless and deterministic strategies. For unbounded cumulative rewards, i.e. when ρ is of the form $\mathtt{C}$, one must additionally identify the maximal end components containing non-zero rewards.

For bounded cumulative rewards ($\rho = \mathtt{C}^{\leqslant k}$) and instantaneous rewards ($\rho = \mathtt{I}^{=k}$) the situation is the same as for bounded probabilistic reachability: the class of deterministic finite-memory strategies are required and a strategy can be synthesised by stepping backwards through the MDP.

Example 5 Returning to our running example we consider strategy synthesis for the query $\mathtt{R}^{moves}_{\min=?}[\,\mathtt{F}\ \mathsf{goal}_2\,]$ where the reward structure *moves* returns 1 for all state-action pairs and all state rewards are zero. This will therefore return a strategy that minimises the expected number of *moves* that the robot needs to make to reach a state satisfying goal_2. We find that the minimum expected number of steps equals $\frac{19}{15}$ and the synthesised memoryless deterministic strategy (not represented in the figure) chooses the actions *south*, *east*, *west* and *north* in s_0, s_1, s_4 and s_5, respectively. ■

3.3.1.3 LTL Properties

We now consider strategy synthesis for a numerical query of the form $\mathtt{P}_{\min=?}[\psi]$ or $\mathtt{P}_{\max=?}[\psi]$ where ψ is an LTL formula. For a given MDP M, the problem can be reduced to the strategy synthesis of a reachability query (see Sect. 3.3.1.1) on the product of M and a deterministic Rabin automaton (DRA) representing ψ [36]. Since for any strategy σ we have:

$$Pr^{\sigma}_{\mathsf{M},\bar{s}}(\psi) \;=\; 1 - Pr^{\sigma}_{\mathsf{M},\bar{s}}(\neg\psi)$$

the problem of finding a minimum probability and strategy for achieving this value can be reduced to finding the maximum probability and corresponding strategy by considering the negated LTL formula. Hence, for the remainder of this section we restrict our attention to the case of maximum numerical queries.

For any LTL formula ψ using atomic propositions from AP, we can construct a DRA A_ψ with alphabet 2^{AP} that represents it [33, 108]. More precisely, we have that an infinite path $\omega = s_0 \xrightarrow{a_0} s_1 \xrightarrow{a_1} s_2 \cdots$ of M satisfies ψ if and only if the infinite word $L(s_0)L(s_1)L(s_2)\ldots$ is in the language of A_ψ. To perform strategy synthesis, we proceed by constructing the product MDP, denoted $\mathsf{M}{\otimes}\mathsf{A}_\psi$, of M and A_ψ. Next we find the maximal end components of this MDP which meet the acceptance conditions of A_ψ and label the states of these components with the atomic proposition acc. This then reduces the problem to a maximum probabilistic reachability query since:

$$Pr^{\max}_{\mathsf{M},\bar{s}}(\psi) = Pr^{\max}_{\mathsf{M}\otimes\mathsf{A}_\psi,(\bar{s},\bar{q})}(\mathtt{F}\ \mathsf{acc})\,.$$

We can now follow the approach described in Sect. 3.3.1.1 to synthesise a *memoryless* deterministic strategy for $\mathsf{M}{\otimes}\mathsf{A}_\psi$ which maximises the probability of reaching accepting end components (and then stays in those end components, visiting each state infinitely often). This strategy can then be used to construct an optimal *finite-memory deterministic* strategy of M for the query $\mathtt{P}_{\max=?}[\psi]$.

Example 6 Returning again to the running example of a robot (Fig. 3.4), we consider synthesising a strategy for the query $\mathtt{P}_{\max=?}[(\mathtt{G}\ \neg\mathsf{hazard}) \wedge (\mathtt{G}\ \mathtt{F}\ \mathsf{goal}_1)]$. This corresponds to finding a strategy which maximises the probability of avoiding a hazard-labelled state *and* visiting a goal_1 state infinitely often. We find that the maximum probability equals 0.1 and that, in this case, a memoryless strategy suffices for achieving optimality. The synthesised strategy selects *south* in state s_0, which leads to state s_4 with probability 0.1. We then remain in states s_4 and s_5 indefinitely by choosing actions *east* and *west*, respectively. ■

3.3.2 *Multi-objective Controller Synthesis*

We now extend the synthesis problem to *multi-objective* queries, this concerns finding a strategy σ that simultaneously satisfies multiple quantitative properties. We first describe the case for LTL properties and then outline how this can be extended.

Definition 3.16 (*Multi-objective probabilistic LTL*) A *multi-objective probabilistic LTL property* is a conjunction $\phi = \mathtt{P}_{\bowtie_1 p_1}[\psi_1] \wedge \ldots \wedge \mathtt{P}_{\bowtie_n p_n}[\psi_n]$ where $\psi_1, \ldots \psi_n$ are LTL formulae and, for $1{\leqslant}i{\leqslant}n$, $\bowtie_i \in \{<, \leqslant, \geqslant, >\}$ and $p_i \in [0, 1]$. For MDP M and strategy σ, we have $\mathsf{M}, \sigma, \bar{s} \models \phi$ if $\mathsf{M}, \sigma, \bar{s} \models \mathtt{P}_{\bowtie_i p_i} \psi_i$ for all $1{\leqslant}i{\leqslant}n$.

The first algorithm for multi-objective probabilistic LTL strategy synthesis was presented in [42]. Here we outline an adapted version of this, based on [45], which uses DRAs. The overall approach is similar to standard (single-objective) LTL strategy synthesis in that it constructs a product automaton and reduces the problem to (multi-objective) reachability.

First, for each $1 \leqslant i \leqslant n$, we can ensure that $\bowtie_i$ is a lower bound ($\geqslant$ or $>$) in each formula $\mathtt{P}_{\bowtie_i p_i} \psi_i$ by negating the formulae ψ_i where necessary. The next step is to build a DRA A_{ψ_i} to represent each LTL formula. Using these automata we then build the product MDP $\mathsf{M}' = \mathsf{M} \otimes \mathsf{A}_{\psi_1} \otimes \cdots \otimes \mathsf{A}_{\psi_n}$. For each combination $X \subseteq \{1, \ldots, n\}$ of objectives we find the end components of M' that are accepting for each of the DRAs in the set $\{\mathsf{A}_i \mid i \in X\}$. A special sink state for X is then added to the product MDP M' for X where for $1 \leqslant i \leqslant n$ we label this sink with acc_i if and only if $i \in X$ and we add transitions from states in the end components found to this sink state. After we have added these components, the problem on M reduces to a multi-objective probabilistic reachability problem on M' of the form $\mathtt{P}_{\bowtie_1 p_1} \mathtt{F}\ \mathsf{acc}_1 \wedge \ldots \wedge \mathtt{P}_{\bowtie_n p_n} \mathtt{F}\ \mathsf{acc}_n$ which can be solved through a linear programming (LP) problem [42], or a value iteration based solution method [46].

The class of strategies required for multi-objective probabilistic LTL is *finite-memory* and *randomised*. A memoryless randomised strategy for the product automaton M' can be obtained, for example, directly from the solution of the LP problem and then, similarly to LTL objectives (in Sect. 3.3.1.3), we can convert this to a finite-memory, randomised strategy for M.

We now summarise several useful extensions and improvements. For details of the algorithms and any restrictions or assumptions that are required see the relevant references.

Boolean combinations of LTL objectives. The approach can be extended to general Boolean combinations of formulae, rather than just conjunctions as presented in Definition 3.16. This is achieved by translating into disjunctive normal form [42, 45].

Expected reward objectives. One can allow unbounded cumulative reward formulae in addition to LTL formulae. The method outlined above has been extended in [45] to include such reward formulae. In addition, an alternative approach, using value iteration, presented in [46], allows bounded cumulative reward formulae. This approach has also been shown to provide significant efficiency gains in practice.

Numerical multi-objective queries. One can again consider numerical queries which return optimal values rather than `true` or `false`. For example, rather than synthesising a strategy satisfying $\mathtt{P}_{\bowtie_1 p_1} \psi_1 \wedge \mathtt{P}_{\bowtie_2 p_2} \psi_2$, we can instead find a strategy that maximises the probability of satisfying the path formula ψ_1, whilst simultaneously satisfying $\mathtt{P}_{\bowtie_2 p_2} \psi_2$. The method outlined above using linear programming is easily extended to handle such numerical queries through the addition of an objective function.

Pareto queries. To analyse the trade-off between multiple objectives we can construct the corresponding *Pareto curve* or an approximation of it [46]. For example, suppose we are interested in maximising the probabilities of two LTL formulae ψ_1 and ψ_2 for the MDP M, then the Pareto curve consists of the bounds $(p_1, p_2) \in [0, 1]^2$ such that:

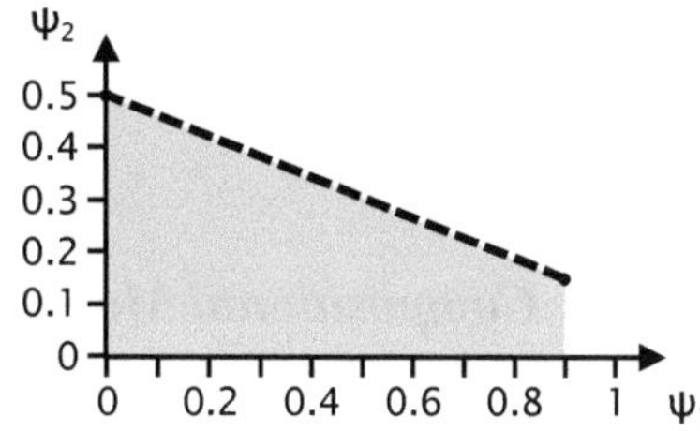

Fig. 3.6 Pareto curve (*dashed line*) for maximisation of the probabilities of LTL formulae $\psi_1 = \mathtt{G}\,\neg\mathsf{hazard}$ and $\psi_2 = \mathtt{G}\,\mathtt{F}\,\mathsf{goal}_1$ (see Example 7)

- there exists a strategy σ such that $Pr^{\sigma}_{\mathsf{M},\bar{s}}(\psi_1)\geqslant p_1$ and $Pr^{\sigma}_{\mathsf{M},\bar{s}}(\psi_2)\geqslant p_2$;
- if either bound p_1 or p_2 is increased, no strategy σ exists satisfying $Pr^{\sigma}_{\mathsf{M},\bar{s}}(\psi_1)\geqslant p_1$ and $Pr^{\sigma}_{\mathsf{M},\bar{s}}(\psi_2)\geqslant p_2$ without decreasing the other bound.

Example 7 We return again to the robot example presented in Fig. 3.4. Recall that, in Example 6, we considered the numerical query $\mathtt{P}_{\max=?}[(\mathtt{G}\,\neg\mathsf{hazard}) \wedge (\mathtt{G}\,\mathtt{F}\,\mathsf{goal}_1)]$ and found that the optimal probability was 0.1. Instead here we consider each conjunct of the LTL formula as a separate objective and, to ease notation, let $\psi_1 = \mathtt{G}\,\neg\mathsf{hazard}$ and $\psi_2 = \mathtt{G}\,\mathtt{F}\,\mathsf{goal}_1$.

Consider the numerical multi-objective query that maximises the probability of satisfying ψ_2 whilst satisfying $\mathtt{P}_{\geqslant 0.7}[\psi_1]$. We find that the optimal value, i.e. the maximum probability for satisfying ψ_2, equals $\frac{41}{180} \approx 0.2278$. The corresponding strategy is randomised and, in state s_0, chooses *east* with probability approximately 0.3226 and *south* with probability approximately 0.6774.

Finally, the Pareto curve for maximising the probabilities of the LTL formulae ψ_1 and ψ_2 is presented in Fig. 3.6. The dashed line in the figure form the Pareto curve, while the grey shaded area below shows all points (x, y) for which there is a strategy satisfying $\mathtt{P}_{\geqslant x}[\psi_1] \wedge \mathtt{P}_{\geqslant y}[\psi_2]$. ■

3.4 Modelling and Verification of Large Probabilistic Systems

In practice, models of real-life systems are often large and complex, and their state space has a tendency to grow exponentially with the size of the system itself, a phenomenon known as the state space explosion problem.

In this section, we discuss some approaches that facilitate both the modelling and verification of large complex probabilistic systems. We first describe higher level modelling of systems comprising multiple components using the notion of parallel composition. Then we describe verification techniques designed to scale up to large, complex systems. Many such approaches exist; examples include symbolic model checking [12, 94], partial order reduction [50], symmetry reduction [39, 77], and bisimulation minimisation [72]. In this presentation, we focus on two particular methods: *compositional verification*, using an assume-guarantee framework, and *abstraction refinement*. We conclude the section with a case study that illustrates both of these techniques.

3.4.1 Compositional Modelling of Probabilistic Systems

Complex system designs usually comprise multiple components operating in parallel. For such a system, if there is probabilistic behaviour present in the system, then an MDP is the natural mathematical model for the system as non-determinism can be used to represent the concurrency between the components. However, for compositional modelling and analysis, probabilistic automata (PAs) [100, 101], a minor generalisation of MDPs, are a more suitable formalism. The key difference is that states of a PA can have more than one transition labelled by the same action.

Definition 3.17 (*Probabilistic automaton*) A *probabilistic automaton* (PA) is a tuple $\mathsf{M} = (S, \bar{s}, A, \delta, L)$, where:

- S is a finite set of states;
- $\bar{s} \in S$ is an initial state;
- A is a finite alphabet;
- $\delta \subseteq S \times (A \cup \{\tau\}) \times Dist(S)$ is a finite probabilistic transition relation;
- $L : S \rightarrow 2^{AP}$ is a state labelling function.

The difference from Definition 3.5 is that we now have a transition relation as opposed to a transition function and allow transitions to be labelled with the silent action τ, representing transitions that are internal to the component being represented.

The notions basic for MDPs presented in Sect. 3.2.2, such as paths and strategies, carry over straightforwardly to PAs. Reward structures for PAs can be defined in exactly the same way as for MDPs (see Definition 3.6), although here a strategy chooses a particular transition (element of δ) as opposed to an action in a state. The semantics of the PRISM logic (see Definition 3.3) and corresponding model checking algorithm presented in Sect. 3.2.2 for MDPs also carry over to PAs.

We next describe parallel composition of PAs, first introduced in [100, 101].

Definition 3.18 (*Parallel composition of PAs*) If $\mathsf{M}_i = (S_i, \bar{s}_i, A_i, \delta_i, L_i)$ are PAs for $i = 1, 2$, then their *parallel composition* $\mathsf{M}_1 \| \mathsf{M}_2 = (S_1 \times S_2, (\bar{s}_1, \bar{s}_2), A_1 \cup A_2, \delta, L)$ is the PA where $((s_1, s_2), a, \mu) \in \delta$ if and only if one of the following holds:

- $a \in A_1 \cap A_2$, $\mu = \mu_1 \times \mu_2$, $(s_1, a, \mu_1) \in \delta_1$ and $(s_2, a, \mu_2) \in \delta_2$;
- $a \in A_1 \backslash A_2$, $\mu = \mu_1 \times \eta_{s_2}$ and $(s_1, a, \mu_1) \in \delta_1$;
- $a \in A_2 \backslash A_1$, $\mu = \eta_{s_1} \times \mu_2$ and $(s_2, a, \mu_2) \in \delta_2$;

and $L((s_1, s_2)) = L_1(s_1) \cup L_2(s_2)$.

The above definition allows several components to synchronise over the same action, so called multi-way synchronisation, as used by the process algebra CSP [99]. It also assumes that the components M_1 and M_2 synchronise over their common actions

$A_1 \cap A_2$. Definition 3.18 can easily be adapted to use other definitions of synchronisation, such as the two-way synchronisation used by the process algebra CCS [89], or to incorporate additional process algebraic operators for hiding or renaming actions.

Below, we demonstrate how a reward structure for a system can be constructed from the reward structures of its components. In this definition we have used addition as this is used in later case studies, however we can easily use other arithmetic operations depending on the quantities that the reward structure represents.

Definition 3.19 If $\mathsf{M}_i = (S_i, \bar{s}_i, A_i, \delta_i, L_i)$ are PAs with reward structures $r_i = (r_{S_i}, r_{A_i})$ for $i = 1, 2$, then the composed reward structure $r = (r_S, r_A)$ for $\mathsf{M}_1 \| \mathsf{M}_2$ is such that for any $(s_1, s_2) \in S_1 \times S_2$ and $a \in A_1 \cup A_2$:

$$
\begin{aligned}
r_S((s_1, s_2)) &= r_{S_1}(s_1) + r_{S_2}(s_2) \\
r_A((s_1, s_2), a) &= \begin{cases} r_{A_1}(s_1, a) + r_{A_2}(s_2, a) & \text{if } a \in A_1 \cap A_2 \\ r_{A_1}(s_1, a) & \text{if } a \in A_1 \backslash A_2 \\ r_{A_2}(s_2, a) & \text{if } a \in A_2 \backslash A_1 . \end{cases}
\end{aligned}
$$

3.4.2 *Compositional Probabilistic Model Checking*

We now describe an approach for *compositional* verification of probabilistic automata presented in [81], based on the popular *assume-guarantee* paradigm. This allows the verification of complex system to be performed through the analysis of individual components of the system in isolation, rather than verifying the much larger complete system. We begin by defining the underlying concepts and then illustrate two of the assume-guarantee proof rules.

The approach is based on the use of linear-time, *action-based* properties Ψ, which are defined in terms of the actions that label the transitions of a probabilistic automaton (or MDP). This is in contrast to the properties discussed elsewhere in this chapter, which are defined in terms of the atomic propositions that label states.[3] More precisely, a property Ψ represents a set of infinite words over the set A of action labels of a probabilistic automaton M. An infinite path ω of M satisfies Ψ, written $\mathsf{M}, \omega \models \Psi$, if the *trace* of π (the sequence of actions labelling its transitions, ignoring silent τ actions) is in the set of infinite words defining Ψ. Then, following the same style as the other property specifications introduced earlier, the property $\mathrm{P}_{\bowtie p}[\Psi]$ states that, for all strategies of the probabilistic automaton M, the probability of a path satisfying Ψ is within the interval given by $\bowtie p$.

We focus our attention here on compositional verification of a class of linear-time action-based properties called *regular safety properties*.

Definition 3.20 (*Regular safety property*) A *safety property* Ψ_P represents a set of infinite words over an alphabet α which is characterised by a set of 'bad prefixes':

[3] In fact, state and action-labelled variants of temporal logics are equally expressive [90].

finite words of which any extension is *not* in the set. A *regular safety property* is a safety property whose set of bad prefixes can be represented as a regular language.

Probabilistic safety properties are of the form $\mathtt{P}_{\geqslant p}[\Psi_P]$, where Ψ_P is a regular safety property. These can be used to capture a variety of useful properties of probabilistic models, including:

- the probability of no failures occurring is at least 0.99;
- event a always occurs before event b with probability at least 0.75;
- the probability of completing a task within k steps is at least 0.9.

A technical requirement of the compositional verification approach described here is the use of *partial strategies*, which can opt to (with some probability) take none of the available actions and remain in the current state. In [81] it is shown that by considering only *fair* strategies, that is the strategies that choose an action from each component of the system infinitely often, this requirement can be removed. We first define the alphabet extension of a PA.

Definition 3.21 (*Alphabet extension of PA*) For a PA $\mathsf{M} = (S, \bar{s}, A, \delta, L)$ and set of actions α, we *extend* M's alphabet to α, denoted by the PA $\mathsf{M}[\alpha]$, as follows: $\mathsf{M}[\alpha] = (S, \bar{s}, A{\cup}\alpha, \delta', L)$ where $\delta' = \delta{\cup}\{(s, a, \eta_s) \mid s{\in}S \wedge a{\in}\alpha\backslash A\}$.

The approach uses *probabilistic assume-guarantee triples*. These take the form $\langle \mathtt{P}_{\geqslant p_\mathsf{A}}[\Psi_\mathsf{A}]\rangle\, \mathsf{M}\, \langle \mathtt{P}_{\geqslant p_\mathsf{G}}[\Psi_\mathsf{G}]\rangle$ where $\Psi_\mathsf{A}, \Psi_\mathsf{G}$ are regular safety properties (see Definition 3.20) and M is a PA. Informally, the triple means: 'whenever M is part of a system satisfying Ψ_A with probability at least p_A, the system satisfies Ψ_G with probability at least p_G'. Formally, we have the following definition.

Definition 3.22 (*Probabilistic assume-guarantee triple*) If $\Psi_\mathsf{A}, \Psi_\mathsf{G}$ are regular safety properties, $p_\mathsf{A}, p_\mathsf{G} \in [0, 1]$ bounds, $\mathsf{M} = (S, \bar{s}, A, \delta, L)$ is a PA and $\alpha_\mathsf{G} \subseteq \alpha_\mathsf{A} \cup A$, then $\langle \mathtt{P}_{\geqslant p_\mathsf{A}}[\Psi_\mathsf{A}]\rangle\, \mathsf{M}\, \langle \mathtt{P}_{\geqslant p_\mathsf{G}}[\Psi_\mathsf{G}]\rangle$ is a *probabilistic assume-guarantee triple*, meaning:

$$\forall \sigma{\in}\Sigma_{\mathsf{M}[\alpha_\mathsf{A}]} \,.\, \big(\mathsf{M}[\alpha_\mathsf{A}], \sigma, \bar{s} \models \mathtt{P}_{\geqslant p_\mathsf{A}}[\Psi_\mathsf{A}] \rightarrow \mathsf{M}[\alpha_\mathsf{A}], \sigma, \bar{s} \models \mathtt{P}_{\geqslant p_\mathsf{G}}[\Psi_\mathsf{G}]\big)\,.$$

The use of $\mathsf{M}[\alpha_\mathsf{A}]$, i.e. M extended to the alphabet of Ψ_A, in the above is needed to allow the assumption to refer to actions not used in M. Verifying that a triple $\langle \mathtt{P}_{\geqslant p_\mathsf{A}}[\Psi_\mathsf{A}]\rangle\, \mathsf{M}\, \langle \mathtt{P}_{\geqslant p_\mathsf{G}}[\Psi_\mathsf{G}]\rangle$ holds requires the use of multi-objective model checking, as discussed in Sect. 3.3.2, as the following proposition demonstrates.

Proposition 3.1 ([81]) *If* $\Psi_\mathsf{A}, \Psi_\mathsf{G}$ *are regular safety properties,* $p_\mathsf{A}, p_\mathsf{G} \in [0, 1]$ *and* M *is a PA, then* $\langle \mathtt{P}_{\geqslant p_\mathsf{A}}[\Psi_\mathsf{A}]\rangle\, \mathsf{M}\, \langle \mathtt{P}_{\geqslant p_\mathsf{G}}[\Psi_\mathsf{G}]\rangle$ *if and only if*

$$\neg\exists \sigma{\in}\mathsf{M}[\alpha_\mathsf{A}] \,.\, \big(\mathsf{M}[\alpha_\mathsf{A}], \sigma, \bar{s} \models \mathtt{P}_{\geqslant p_\mathsf{A}}[\Psi_\mathsf{A}] \wedge \mathsf{M}[\alpha_\mathsf{A}], \sigma, \bar{s} \not\models \mathtt{P}_{\geqslant p_\mathsf{G}}[\Psi_\mathsf{G}]\big)\,.$$

Based on the definitions given above, [81] presents the following *asymmetric* assume-guarantee proof rule for a two component system $\mathsf{M}_1 \| \mathsf{M}_2$.

Proposition 3.2 ([81]) *If Ψ_A, Ψ_G are regular safety properties, p_A, $p_\mathsf{G} \in [0,1]$ and $\mathsf{M}_1, \mathsf{M}_2$ are PAs such that $\alpha_\mathsf{A} \subseteq A_1$ and $\alpha_\mathsf{G} \subseteq A_2 \cup \alpha_\mathsf{A}$, then the following proof rule holds:*

$$\frac{\mathsf{M}_1, \bar{s}_1 \models \mathtt{P}_{\geqslant p_\mathsf{A}}[\Psi_\mathsf{A}] \qquad \langle \mathtt{P}_{\geqslant p_\mathsf{A}}[\Psi_\mathsf{A}]\rangle\, \mathsf{M}_2\, \langle \mathtt{P}_{\geqslant p_\mathsf{G}}[\Psi_\mathsf{G}]\rangle}{\mathsf{M}_1 \parallel \mathsf{M}_2, (\bar{s}_1, \bar{s}_2) \models \mathtt{P}_{\geqslant p_\mathsf{G}}[\Psi_\mathsf{G}]} \quad \text{(ASYM)}$$

Proposition 3.2 presents a method to verify the property $\mathtt{P}_{\geqslant p_\mathsf{G}}[\Psi_\mathsf{G}]$ on $\mathsf{M}_1 \| \mathsf{M}_2$ in a *compositional* fashion. More precisely, verification reduces to two sub-problems, one for each premise of the rule:

1. computing the optimal probability of a regular safety property on M_1;
2. performing multi-objective model checking on $\mathsf{M}_2[\alpha_\mathsf{A}]$.

A limitation of the above rule is the fact it is asymmetric: we analyse the component M_2 using an assumption about the component M_1, but when verifying M_1 we cannot make any assumptions about M_2. Below, we give a proof rule which does allow the use of additional assumptions in this way.

Proposition 3.3 ([81]) *If $\Psi_{\mathsf{A}_1}, \Psi_{\mathsf{A}_2}, \Psi_\mathsf{G}$ are regular safety properties, $p_{\mathsf{A}_1}, p_{\mathsf{A}_2}, p_\mathsf{G} \in [0,1]$ and $\mathsf{M}_1, \mathsf{M}_2$ are PAs such that $\alpha_{\mathsf{A}_2} \subseteq A_2$, $\alpha_{\mathsf{A}_1} \subseteq A_2 \cup \alpha_{\mathsf{A}_2}$ and $\alpha_\mathsf{G} \subseteq A_1 \cup \alpha_{\mathsf{A}_1}$, then the following proof rule holds:*

$$\frac{\mathsf{M}_2, \bar{s}_2 \models \mathtt{P}_{\geqslant p_{\mathsf{A}_2}}[\Psi_{\mathsf{A}_2}] \qquad \langle \mathtt{P}_{\geqslant p_{\mathsf{A}_2}}[\Psi_{\mathsf{A}_2}]\rangle\, \mathsf{M}_1\, \langle \mathtt{P}_{\geqslant p_{\mathsf{A}_1}}[\Psi_{\mathsf{A}_1}]\rangle \qquad \langle \mathtt{P}_{\geqslant p_{\mathsf{A}_1}}[\Psi_{\mathsf{A}_1}]\rangle\, \mathsf{M}_2\, \langle \mathtt{P}_{\geqslant p_\mathsf{G}}[\Psi_\mathsf{G}]\rangle}{\mathsf{M}_1 \parallel \mathsf{M}_2, (\bar{s}_1, \bar{s}_2) \models \mathtt{P}_{\geqslant p_\mathsf{G}}[\Psi_\mathsf{G}]} \quad \text{(CIRC)}$$

For further details of the assume-guarantee proof rules, including extensions to allow both ω-regular properties and reward-based properties, see [81].

3.4.3 *Quantitative Abstraction Refinement*

An alternative way to verify large, complex systems is using *abstraction-refinement* techniques, which have been established as one of the most effective ways of performing *non-probabilistic* model checking on complex systems [30]. The basic idea is to build a small abstract model, by removing details of the complex concrete system which are not relevant to the property of interest, which is consequently easier to analyse. The abstract model is constructed in such a way that, when the property of interest is verified `true` for the abstraction, the property also holds for the concrete system. On the other hand, if the property does not hold for the abstraction, then information from the model checking process (typically a counterexample) is used either to show that the property is false in the concrete system or to refine the abstraction. This process forms the basis of a loop which refines the abstraction until the property is shown either to be `true` or `false` in the concrete system.

In the case of probabilistic model checking a number of abstraction-refinement approaches exist. D'Argenio et al. [34] introduce an approach for verifying reachability properties of PAs based on probabilistic simulation [100] and implement a corresponding tool RAPTURE [66]. Properties are analysed on abstractions obtained through successive refinements, starting from a coarse partition derived from the property under study. This approach only produces lower (upper) bounds for the minimum (maximum) reachability probabilities. Based on [34] and using predicate abstraction [48], an abstraction-refinement framework for PAs is developed in [63, 110] and implemented in the PASS tool [53]. The framework is used for verifying or refuting properties of the form 'the maximum probability of error is at most p' for a given probability threshold p. Since abstractions produce only upper bounds on maximum probabilities, to refute a property, probabilistic counterexamples [58] (comprising multiple paths whose combined probability exceeds p) are generated. If these paths are spurious, then they are used to generate further predicates using interpolation.

An alternative framework is presented in [75] where the key idea is to maintain a distinction between the non-determinism from the original PA and the non-determinism introduced during the abstraction process. To achieve this, abstractions of PAs are modelled as stochastic two player games (see Sect. 3.2.3), where the two players correspond to the two different forms of non-determinism. Analysis of these abstract models results in a separate lower and upper bound for the property of interest (e.g. an optimal reachability probability or expected reward value). These bounds both provide quantitative measure of the quality (or preciseness) of the abstraction and an indication of how to improve it. The abstraction-refinement framework is presented in Fig. 3.7. The framework starts with a simple, coarse abstraction (i.e. partition of the state space) and then refines the abstraction until the difference between the bounds is below some threshold value ε. Two methods for automatically refining abstractions are considered. The first is based on the difference between specific strategies that achieve the lower and upper bounds. The second method differs by considering all the strategies that achieve the bounds. In [74, 79], this game-based abstraction and refinement framework has been used to develop verification techniques for probabilistic software and probabilistic timed automata (see Sect. 3.5.1), respectively.

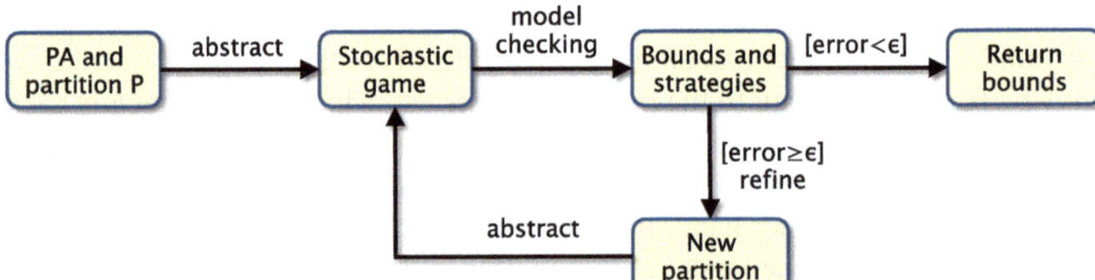

Fig. 3.7 Quantitative abstraction-refinement framework for PAs [75]

3.4.4 Case Study: The Zeroconf Protocol

This case study concerns the ZeroConf dynamic configuration protocol for IPv4 link-local addresses [29] used to enable devices to connect to a local network. The protocol is a distributed 'plug-and-play' solution to IP address configuration. This case study was originally introduced and analysed using probabilistic model checking in [82]. The compositional approach present in Sect. 3.4.2 and abstraction-refinement framework present in Sect. 3.4.3 have since been used to analyse this model in [75, 81], respectively.

The protocol is used to configure an IP address for a device when it first joins a local network. This address is then used for the communication between the device and others within the network. When connecting to the network, a device first randomly selects an address from the 65,024 possible local IP addresses. The device then waits a random time of between 0 and 2 s before sending a number of probes including the chosen address over 4 second intervals. These probes are sent all of the other hosts of the network and are used to check whether any other device is using the chosen address. If the original device gets a message back saying the address is already in use it will restart the protocol by reconfiguring. If the host repeats this process 10 times, it 'backs off' and remains idle for at least one minute. If the host does not get a reply to any of the probes it commences to use the chosen IP address. We assume that messages can also get lost with a fixed probability.

The model of this system studied in [81] consists of the parallel composition of two PAs: one automaton representing a new device joining the local network and the other representing the environment, i.e. the existing network including the devices present in the network. Using the composition approach, [81] analysed the following properties:

- the minimum probability that the new host employs a fresh IP address;
- the minimum probability that the new host is configured by time T;
- the minimum probability that the protocol terminates;
- the minimum and maximum expected time for the protocol to terminate.

The first two can be expressed using regular safety properties and were verified compositionally by applying the rules presented in Propositions 3.2 and 3.3. In the case of the final two properties, extensions of the rules to LTL and reward properties were used.

The case study developed in [75] using the game based abstraction-refinement described in Sect. 3.4.3 also considers a new device joining a network. The network consists of N devices and there are M available addresses. The model is the parallel composition of $2{\cdot}N{+}1$ component PAs: the device joining the network and N pairs of channels for the two-way communication between the new device and each of the configured devices. Figure 3.8 presents, for a fixed abstraction, the upper and lower bounds obtained when calculating the maximum probability that the new device has not configured successfully by time T. The figure also includes the results obtained when model checking the concrete system. The graphs demonstrate how the differences between the lower and upper bounds can be used to quantify the utility of

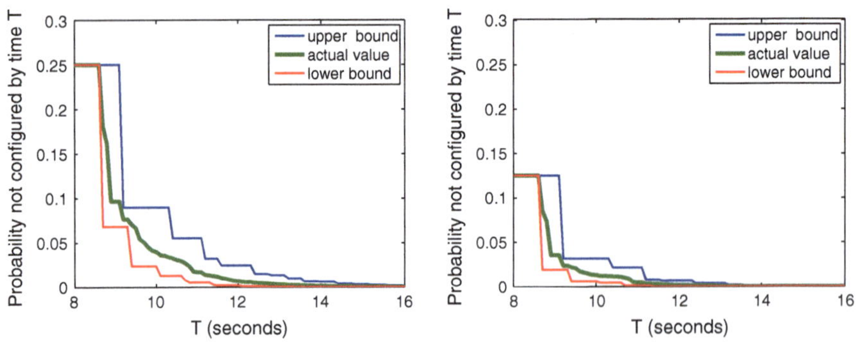

(a) N=8 and M=32

(b) N=8 and M=64

Fig. 3.8 Zeroconf case study: maximum probability device not configured successfully by T [75]

the abstraction. For the fixed abstraction, the number of states in the abstraction is independent of M and equals 881, on the other hand, the concrete system has 432,185 states when $M = 32$ and 838,905 states when $M = 64$.

3.5 Real-Time Probabilistic Model Checking

So far, we have seen *discrete-time* models which exhibit just probabilistic behaviour (DTMCs) or both probabilistic and non-deterministic behaviour (MDPs and SMGs). However, it is also often important to model the *real-time* characteristics and the interplay between the different types of behaviour. Relevant application areas range from wireless communication, automotive networks to security protocols. We will first give an overview of *probabilistic timed automata*, a formalism that allows for the modelling of systems exhibiting non-deterministic, probabilistic and real-time behaviour and a case study using this formalism. The final part of this section concerns *continuous-time Markov chains*, which are also suitable for modelling systems with probabilistic and real-time characteristics.

3.5.1 Probabilistic Timed Automata

Probabilistic timed automata (PTAs) [17, 68, 84] extend classical timed automata [5] with discrete probabilistic choice. Real-time behaviour is modelled through *clocks* which are variables whose values range over the non-negative reals and increase at the same rate as time. For the remainder of this section, we assume a finite set of clocks $\mathscr{X}$. Before we give the formal definition of PTAs we require the following notation and definitions relating to clocks.

A function $v : \mathscr{X} \to \mathbb{R}_{\geqslant 0}$ is called a *clock valuation* and we denote the set of all clock valuations by $\mathbb{R}_{\geqslant 0}^{\mathscr{X}}$. For a clock valuation $v \in \mathbb{R}_{\geqslant 0}^{\mathscr{X}}$, real-time delay $t \in \mathbb{R}_{\geqslant 0}$ and set of clocks $X \subseteq \mathscr{X}$, we use $v+t$ to denote the clock valuation obtained from v by incrementing all clock values by t and $v[X:=0]$ for the clock valuation obtained from v by resetting the clocks in X to 0. We let $\mathbf{0}$ denote the clock valuation that assigns 0 to all clocks in $\mathscr{X}$. We define the set of *clock constraints* over $\mathscr{X}$, denoted $CC(\mathscr{X})$, by the syntax:

$$\zeta ::= \texttt{true} \mid x \leqslant d \mid c \leqslant x \mid x+c \leqslant y+d \mid \neg\zeta \mid \zeta \wedge \zeta$$

where $x, y \in \mathscr{X}$ and $c, d \in \mathbb{N}$. A clock valuation v satisfies a clock constraint ζ, denoted $v \models \zeta$, if the constraint ζ resolves to $\texttt{true}$ after substituting the occurrences of each clock x with $v(x)$.

Definition 3.23 (*PTA syntax*) A *probabilistic timed automaton* (PTA) is a tuple of the form $\mathsf{P} = (L, \bar{l}, \mathscr{X}, Act, inv, enab, prob, L_{\mathsf{P}})$ where:

- L is a finite set of *locations* and $\bar{l} \in L$ is an *initial location*;
- $\mathscr{X}$ is a finite set of *clocks*;
- Act is a finite set of *actions*;
- $inv : L \to CC(\mathscr{X})$ is an *invariant condition*;
- $enab : L \times Act \to CC(\mathscr{X})$ is an *enabling condition*;
- $prob : L \times Act \to Dist(2^{\mathscr{X}} \times L)$ is a (partial) *probabilistic transition function*;
- $L_{\mathsf{P}} : L \to 2^{AP}$ is a labelling function.

A *state* of a PTA is a pair $(l, v) \in L \times \mathbb{R}_{\geqslant 0}^{\mathscr{X}}$ such that $v \models inv(l)$. In any state (l, v), there is a non-deterministic choice between either a certain amount of time elapsing, or an action being performed. If time elapses, then the choice of time $t \in \mathbb{R}_{\geqslant 0}$ requires that the invariant $inv(l)$ remains continuously satisfied while time passes. The resulting state after this transition is $(l, v+t)$. In the case where an action is performed, an action a can only be chosen if it is *enabled*, i.e. if the clock constraint $enab(l, a)$ is satisfied by v. Once an enabled action a is chosen, a set of clocks to reset and a successor location are selected at random, according to the distribution $prob(l, a)$.

Definition 3.24 (*PTA semantics*) Let $\mathsf{P} = (L, \bar{l}, \mathscr{X}, Act, inv, enab, prob, L_{\mathsf{P}})$ be a PTA. The semantics of P is defined as the (infinite-state) MDP $[\![\mathsf{P}]\!] = (S, \bar{s}, A, \delta, L)$ where:

- $S = \{(l, v) \in L \times \mathbb{R}_{\geqslant 0}^{\mathscr{X}} \mid v \models inv(l)\}$;
- $\bar{s} = (\bar{l}, \mathbf{0})$;
- $A = \mathbb{R}_{\geqslant 0} \cup Act$;
- for any $(l, v) \in S$ and $a \in Act \cup \mathbb{R}_{\geqslant 0}$, we have $\delta((l, v), a) = \lambda$ if and only if either:
  - *(time transitions)* $a \in \mathbb{R}_{\geqslant 0}$, $v+t' \models inv(l)$ for all $0 \leqslant t' \leqslant a$, and $\lambda = \eta_{(l, v+a)}$;
  - *(action transitions)* $a \in Act$, $v \models enab(l, a)$ and for each $(l', v') \in S$:

$$\lambda(l', v') = \sum \{\!| \, prob(l, a)(X, l') \mid X \in 2^{\mathscr{X}} \wedge v' = v[X:=0] \, |\!\} ,$$

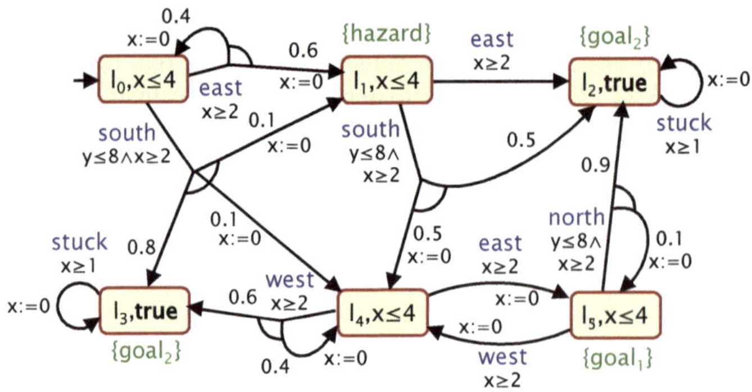

Fig. 3.9 Running example: a PTA P representing a robot moving about a 3×2 grid

- for any $(l, v) \in S$ we have $L(l, v) = L_{\mathsf{P}}(l) \cup \{\zeta \mid \zeta \in CC(\mathscr{X}) \wedge v \models \zeta\}$.

The set of atomic propositions of $[\![\mathsf{P}]\!]$ is the union of the atomic propositions used for labelling the locations L and the clock constraints obtained from the clocks $\mathscr{X}$. We now return to our running example of a robot and extend our MDP model to exhibit real-time behaviour.

Example 8 Figure 3.9 shows a PTA model of our robot moving through terrain that is divided up into a 3×2 grid, extending the MDP model described in Example 3. The PTA has two clocks x and y and each grid section is represented as a location (with initial location l_0). In each location, one or more actions from the set $Act = \{north, east, south, west, stuck\}$ are again available. As before, due to the presence of obstacles, certain directions are unavailable in some states or probabilistically move the robot to an alternative state.

The invariant $x \leqslant 4$ in the locations l_0, l_1, l_4 and l_5 and the fact that the clock x is reset on all transitions entering these locations implies that at most 4 time units can be spent in these locations. While the inclusion of the constraint $x \geqslant 2$ in all guards of the transitions leaving these locations, implies that the robot must remain in these location for at least 2 time units. In addition, the inclusion of $y \leqslant 8$ in the guards on the transitions labelled *north* and *south* and the fact the clock y is never reset, means that the robot can only move 'north' and 'south' during the first 8 time units of operation. ■

As for DTMCs and MDPs (see Sect. 3.2), we can define *reward structures* for PTAs.

Definition 3.25 (*PTA reward structure*) For PTA P with locations L and actions Act, a *reward structure* is given by a pair $r = (r_L, r_{Act})$ where:

- $r_L : L \to \mathbb{R}_{\geqslant 0}$ is a function assigning to each location the rate at which rewards are accumulated as time passes in that location;
- $r_{Act} : L \times Act \to \mathbb{R}_{\geqslant 0}$ is a function assigning the reward of executing each action in each location.

The location rewards of a PTA assign the rate at which rewards are accumulated as time passes in a state, and therefore the corresponding reward structure of the MDP $[\![\mathsf{P}]\!]$ consists of only action rewards. More precisely, in the corresponding reward structure of $[\![\mathsf{P}]\!]$ we have $r_A((l,v),t) = r_L(l){\cdot}t$ and $r_A((l,v),a) = r_{Act}(l,a)$ for all $(l,v) \in L \times \mathbb{R}_{\geqslant 0}^{\mathscr{X}}$, $t \in \mathbb{R}_{\geqslant 0}$ and $a \in Act$. PTAs equipped with reward structures are a probabilistic extension of linearly-priced timed automata (also known as weighted timed automata) [8, 19]. Parallel composition (see Sect. 3.4) can also be extended to PTAs [92] under the restriction that the sets of clocks of the component PTAs are disjoint.

An important issue with regard to the analysis of models exhibiting real-time behaviour is that of *time divergence*. More precisely, we do not consider executions in which time does not advance beyond a certain point. These can be ignored on the grounds that they do not correspond to actual, realisable behaviour of the system being modelled [3, 7]. For a PTA P this corresponds to restricting the analysis of the MDP $[\![\mathsf{P}]\!]$ to the class of time-divergent (or non-Zeno) strategies (those strategies for which the probability of time passing beyond any bound is 1). This clearly has an impact on the complexity of any analysis. However, there are syntactic conditions, derived from analogous results on timed automata [106, 107], which guarantee that all strategies will be time-divergent, see [92] for further details.

The PRISM logic (see Definition 3.3) previously used for specifying properties for DTMCs and MDPs can also be applied to PTAs. There is one key difference since time is now dense as opposed to discrete: the bounds appearing in formulae correspond to bounds on the elapsed time as opposed to bounds on the number of discrete steps. More precisely, path formula $\psi_1 \ \mathtt{U}^{\leqslant k} \ \psi_2$ holds if a state satisfying ψ_2 is reached before k time units have elapsed and, up until that point in time, ψ_1 is continuously satisfied, and the reward formulae $\mathtt{I}^{=k}$ and $\mathtt{C}^{\leqslant k}$ represent the reward at time instant k and the reward accumulated up until k time units have elapsed. As the underlying semantics of a PTA is an MDP the semantics of the logic for PTAs is as given in Definition 3.9, modified for bounded properties due to the different interpretation for PTAs given above [92]. The logic can also be extended to allow more general timing properties through *formula clocks* and *freeze quantifiers* [84].

There are a number of different model checking approaches for PTAs which support different classes of properties. Each is based on first constructing a finite state MDP and then analysing this MDP (either computing the optimal probability of a path formula or the expected value of a reward formula). Approaches for model checking PTAs include:

- the region graph construction [84];
- the boundary region graph [70];
- the digital clocks method [82];
- forwards reachability [84];
- backwards reachability [85];
- abstraction refinement with stochastic games [79].

For a discussion of the advantages and disadvantages of these approaches, see [92].

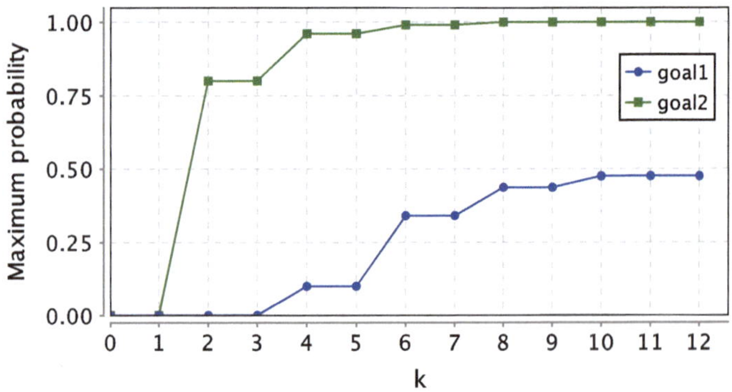

Fig. 3.10 Running example: Time-bounded probabilistic reachability for the PTA model of the robot

Example 9 For the PTA model of the robot given in Example 8 and Fig. 3.9 the maximum probability of reaching a goal_1 labelled state ($\mathtt{P}_{\max=?}[\mathtt{F}\ \mathsf{goal}_1]$) is now 0.4744 as opposed to 0.5 for the MDP model (see Example 3). This is due to the fact that the *north* and *south* actions are only available during the first 8 time units of operation and the robot must remain in the locations l_0, l_1, l_4 and l_5 for between 2 and 4 time units. The minimum expected time to reach a goal_2 state equals 2.5333 and is obtained through the query $\mathtt{R}^{r}_{\min=?}[\ \mathtt{F}\ \mathsf{goal}_2\ ]$ where the reward structure $r = (r_L, r_{Act})$ is such that $r(l) = 1$ and $r(l, a) = 0$ for for all locations l and actions a. Finally, Fig. 3.10 plots results for the time-bounded maximum reachability properties $\mathtt{P}_{\max=?}[\mathtt{F}^{\leqslant k}\ \mathsf{goal}_1]$ and $\mathtt{P}_{\max=?}[\mathtt{F}^{\leqslant k}\ \mathsf{goal}_2]$ as the time bound k varies. ■

Extensions to PTAs. One way of extending PTAs is to allow more general continuous dynamics to model hybrid systems (see Sect. 3.7). We also mention the introduction of continuously-distributed time delays, see for example [2, 83, 88] and probabilistic timed games (see for example [9, 70]), which can build on the success of (non-probabilistic) timed games for the analysis of synthesis problems [18].

3.5.1.1 Case Study: Processor Task Scheduling

This PTA case study is taken from [92] and is based on the *task-graph scheduling* problem described in [22] using (non-probabilistic) timed automata. The case study concerns determining optimal schedulers for either the (expected) time or energy consumption required to compute the arithmetic expression $D \times (C \times (A + B)) + ((A + B) + (C \times D))$ using two processors (P_1 and P_2) that have different speed and energy requirements. Figure 3.11 presents a task graph for computing this expression and shows both the tasks that need to be performed (the subterms of the expression) and the dependencies between the tasks (the order the tasks must be evaluated in).

The specification of the processors, as given in [22], is as follows:

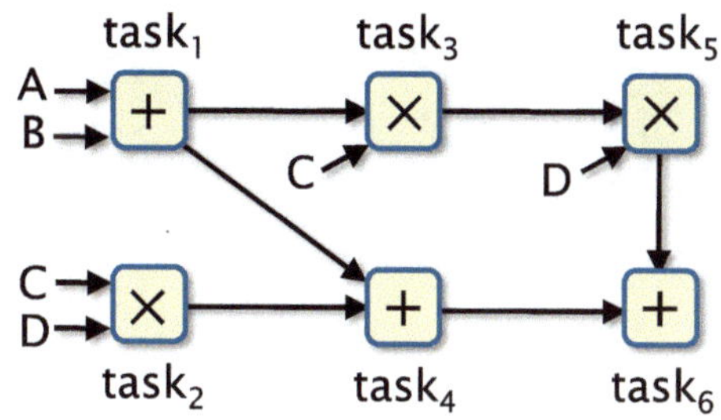

Fig. 3.11 Processor task scheduling problem: computing $D \times (C \times (A + B)) + ((A + B) + (C \times D))$

- *time for addition:* 2 and 5 picoseconds for processors P_1 and P_2;
- *time for multiplication:* 3 and 7 picoseconds for processors P_1 and P_2;
- *idle energy usage:* 10 and 20 Watts for processors P_1 and P_2;
- *active energy usage:* 90 and 30 Watts for processors P_1 and P_2.

The system is formed as the parallel composition of three PTAs—one for each processor and one for the scheduler. In Fig. 3.12a we give a timed automaton representing P_1. The labels *p1_add* and *p1_mult* on the transitions represent an addition and multiplication task being scheduled on P_1, respectively, while the label *p1_done* indicates that the current task has been completed. The PTA includes a clock x which is used to keep track of the time that a task has been running and is therefore reset when a task starts and the invariants and guards correspond to the time required to complete the tasks of addition and multiplication for P_1. The reward structure for computing the expected energy consumption associates a reward of 10 with the *stdby* location and reward 90 with the locations *add* and *mult* (corresponding to the energy usage of process P_1 when idle and active, respectively) and all action rewards are 0. The PTA and reward structure for processor P_2 are similar except for the names of the labels, invariants, guards and reward values correspond to the specification of P_2. After forming the parallel composition, the reward structure for the expected energy consumption then includes the addition of the reward structures for energy consumption of P_1 and P_2. The reward structure for computing the expected time associates a reward of 1 with all locations of the composed system.

In [92] the model of [22] is extended in the following ways.

- A third processor P_3 that has faulty behaviour is added to the system. We assume the faulty processor consumes the same energy consumption as P_2, but is faster (addition takes 3 picoseconds and multiplication 5 picoseconds) and has probability p of failing to successfully complete a task. The PTA model of the P_3 is given in Fig. 3.12b.
- The processors P_1 and P_2 are changed to have random execution times. We assume that, if the original time to perform a task was t, then the time taken is now uniformly distributed between the delays $t-1$, t and $t+1$. Figure 3.12c presents the resulting PTA model of P_1.

For each model, we synthesise the optimal schedulers for both the expected time and energy usage to complete all tasks. To achieve this we used the numerical reward

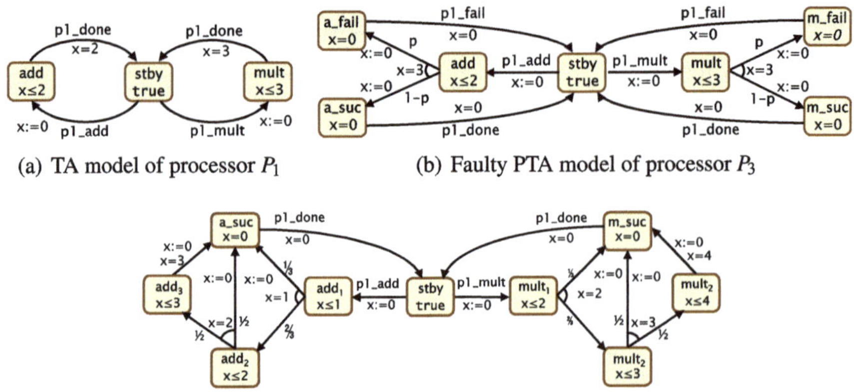

(a) TA model of processor P_1

(b) Faulty PTA model of processor P_3

(c) PTA model P_1 with random delays

Fig. 3.12 PTAs for the task-graph scheduling case study

queries $\mathrm{R}^{time}_{\mathrm{min}=?}[\,\mathtt{F}\ \mathsf{complete}\,]$ and $\mathrm{R}^{energy}_{\mathrm{min}=?}[\,\mathtt{F}\ \mathsf{complete}\,]$ with the reward structures described above.

Basic model. For the basic (non-probabilistic) model, as proposed in [22], an optimal scheduler for minimising the elapsed time to complete all tasks, takes 12 picoseconds to complete all tasks and schedules the tasks as follows:

time	1	2	3	4	5	6	7	8	9	10	11	12	13	14	15	16	17	18	19	20
P_1	$task_1$		$task_3$			$task_5$			$task_4$		$task_6$									
P_2	$task_2$																			

When considering the energy consumption to complete all tasks, an optimal scheduler makes the following choices:

time	1	2	3	4	5	6	7	8	9	10	11	12	13	14	15	16	17	18	19	20
P_1	$task_1$		$task_3$			$task_4$														
P_2	$task_2$							$task_5$							$task_6$					

The above scheduler requires 1.3200 nanojoules and 19 picoseconds to complete all tasks. Since processor P_1 consumes additional energy when active, the first scheduler described, optimising the time to complete all tasks, requires 1.3900 nanojoules.

Faulty processor. When adding the faulty processor P_3 we find that for small values of p (the probability of P_3 failing to successfully complete a task), as P_3 has better performance than P_2, both the optimal expected time and energy consumption can be improved using P_3. However, as the probability of failure increases, P_3's better performance is outweighed by the chance of its failure and using it no longer yields optimal values. For example, below, we give an optimal scheduler for minimising the expected time when $p = 0.25$ which takes 11.0625 picoseconds (the optimal expected time is 12 picoseconds when P_3 is not used). The dark boxes are used to denote the cases when P_3 is scheduled to complete a task, but experiences a fault and does not complete the scheduled task correctly.

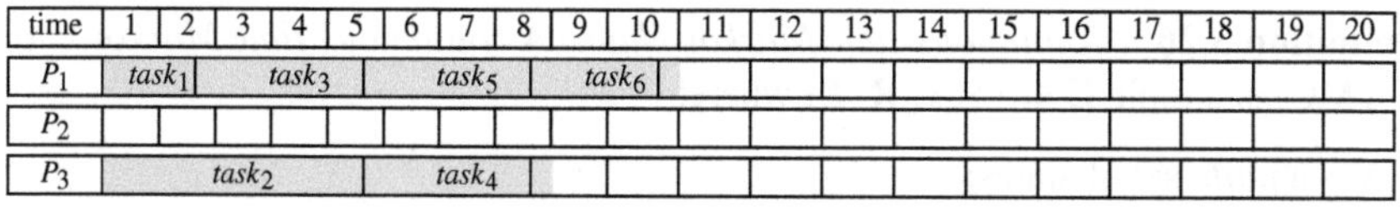

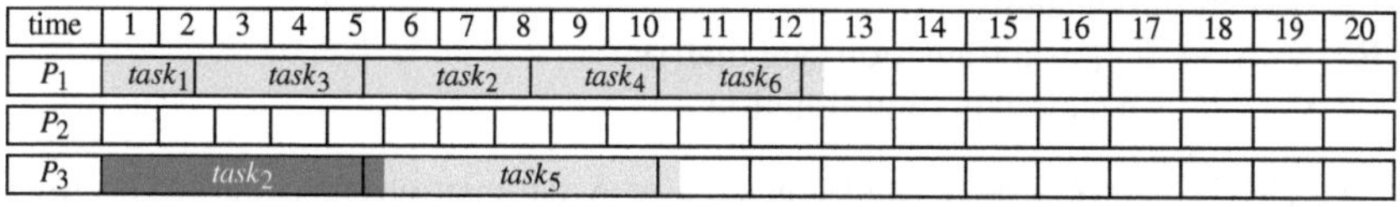

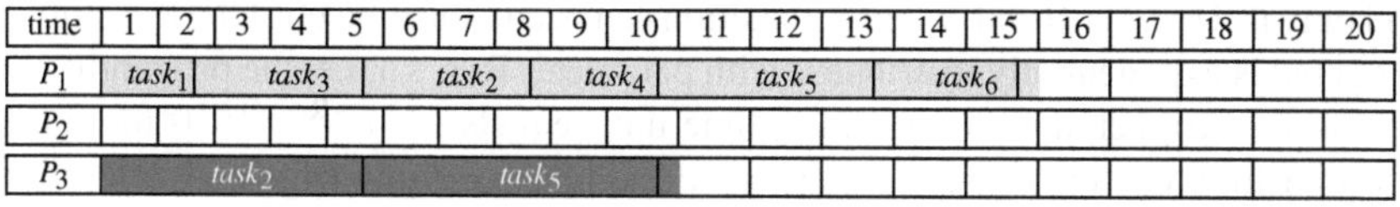

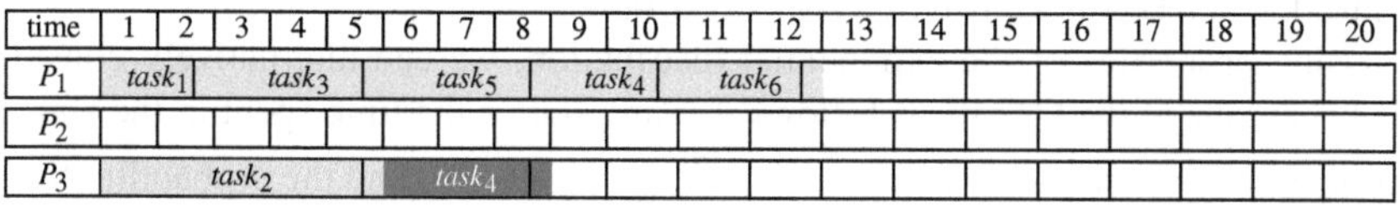

This optimal scheduler uses the processor P_3 for $task_2$ and, if this task is completed successfully, it then uses P_3 for $task_4$. However, if the processor fails to complete $task_2$, P_3 is instead then used for $task_5$ with $task_4$ being rescheduled on P_1.

Random execution times. For this model, the optimal expected time and energy consumption are 12.226 picoseconds and 1.3201 nanojoules, respectively. The optimal schedulers change their decision based upon the delays of previously completed tasks. For example, a scheduler that optimises the elapsed time starts by following the choices for the optimal scheduler described for the basic model: first scheduling $task_1$ followed by $task_3$ on P_1 and $task_2$ on P_2. Due to the random execution times it is now possible for $task_2$ to complete before $task_3$ (if the execution times for $task_1$, $task_2$ and $task_3$ are 3, 6 and 4, respectively) and in this case the optimal decision differs from those made for the basic model. To illustrate this we give one possible set of execution times for the tasks and a corresponding optimal scheduling.

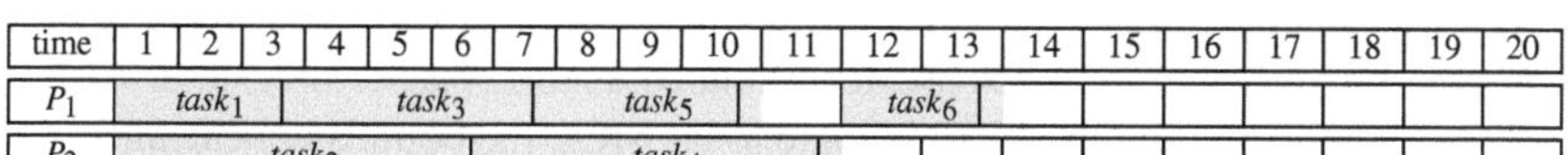

3.5.2 Continuous-Time Markov Chains

Continuous-time Markov chains (CTMCs) are an alternative way to model systems exhibiting probabilistic and real-time behaviour. This model type is very frequently used in performance analysis and can be considered as a real-time extension of DTMCs. While each transition between states in a DTMC corresponds to a discrete time-step, in a CTMC transitions occur in real time.

Definition 3.26 (*Continuous-time Markov chain*) A *continuous-time Markov chain* (CTMC) is a tuple $\mathsf{C} = (S, \bar{s}, \mathbf{R}, L)$ where:

- S is a *finite* set of states;
- $\bar{s} \in S$ is an initial state;
- $\mathbf{R} : S \times S \to \mathbb{R}_{\geqslant 0}$ is a transition rate matrix;
- $L : S \to 2^{AP}$ is a state labelling function.

For a CTMC $\mathsf{C} = (S, \bar{s}, \mathbf{R}, L)$ and states $s, s' \in S$, a transition can occur from s to s' if and only if $\mathbf{R}(s, s')>0$ and, when a transition can occur, the time until the transition is triggered is exponentially distributed with parameter $\mathbf{R}(s, s')$, i.e. the probability the transition is triggered within $t \in \mathbb{R}_{\geqslant 0}$ time-units equals $1 - e^{-\mathbf{R}(s,s')\cdot t}$. If more than one transition can occur from a state, then the first transition triggered determines the next state. This is known as a *race condition*.

Using properties of the exponential distribution, we can alternatively consider the behaviour of the CTMC as follows: for any state s the time spent in the state is exponentially distributed with rate $E(s) \overset{\text{def}}{=} \sum_{s' \in S} \mathbf{R}(s, s')$ and the probability that a transition to state s' is then taken equals $\mathbf{R}(s, s')/E(s)$.

As for DTMCs and MDPs, an execution of a CTMC is represented as a path. However, here, we must also consider the time at which a transition is taken. Formally, a path of a CTMC is a (finite or infinite) sequence $\pi = s_0 t_0 s_1 t_1 s_2 t_2 \ldots$ such that $\mathbf{R}(s_i, s_{i+1})>0$ and $t_i \in \mathbb{R}_{>0}$ for all $i \geqslant 0$. Furthermore, let $time(\pi, i)$ denote the time spent in the $(i+1)$th state, that is t_i.

To define a probability measure over infinite paths of a CTMC, we need to extend the cylinder sets used in the probability measure construction for DTMC (see Sect. 3.2.1) to include time intervals. More precisely, if $s_0, \ldots, s_n$ is a sequence of states such that $\mathbf{R}(s_i, s_{i+1})>0$ for all $0 \leqslant i < n$ and $I_0, \ldots, I_{n-1}$ are non-empty intervals in $\mathbb{R}_{\geqslant 0}$, then the cylinder set $C(s_0, I_0, \ldots, I_{n-1}, s_n)$ is the set of infinite paths such that $\pi \in C(s_0, I_0, \ldots, I_{n-1}, s_n)$ if and only if $\pi(i) = s_i$ and $time(\pi, i) \in I_j$ for all $0 \leqslant i \leqslant n$ and $0 \leqslant j < n$. We can then construct a probability measure Pr_{C,s_0} over the infinite paths of the CTMC. For further details on this construction see [14].

Reward structures can be defined for a CTMC and, as for PTAs, state rewards assign the rate at which rewards are accumulated as time passes in a state. Also, as for PTAs, when applying the PRISM logic to CTMCs, the bounds appearing in path and reward formulae correspond to the elapsed time as opposed to the number of steps performed. It follows that the only difference between model checking DTMCs and CTMCs concerns the analysis of bounded properties. The standard approach for verifying such time-bounded properties is to use uniformisation [49, 67]. For more details on the model checking algorithms for CTMCs see, for example, [14, 78].

To express non-deterministic behaviour, CTMCs can be extended to *continuous-time Markov decision processes* and related models such as *interactive Markov chains* [62] and *Markov automata* [41]. For such models the main difference from model checking MDPs is again when verifying bounded properties which is considerable more complex in the continuous-time setting where the bounds correspond to elapsed time as opposed to the number of discrete steps. Model checking algorithms for such models have been developed, see for example [25], as well as temporal logics which allow specification of more expressive timing requirements; see for example [40].

3.6 Parametric Probabilistic Model Checking

In this section, we consider another extension to the basic technique of probabilistic model checking which provides *parametric* techniques for analysing models. One or more values in definition of the model (for example, a transition probability) or in the property to be verified (for example, a time bound) are provided as a parameter to the verification problem, rather than being instantiated to a specific value. For a numerical query, *parametric model checking* can compute a symbolic expression for the result, as a function of the parameters, rather than a concrete value. For Boolean-valued queries, *parameter synthesis* can be applied to determine the set of all parameter values for which the model is true.

We first consider the parametric model checking of DTMC models and, following this, consider approaches for other probabilistic models.

3.6.1 Parametric Model Checking for DTMCs

Parametric model checking of DTMCs was first proposed by Daws [35] for the logic PCTL. The basic idea is to represent transition probabilities as rational functions and then use a language-theoretic approach to compute the probability of reaching a set of target states. This is done by treating the transition probabilities as letters of an alphabet, converting the DTMC to a finite automaton over this alphabet and then using the state elimination method to determine a rational function representing the probability of reaching the target.

Since the approach of [35] was first presented, a variety of extensions and implementations have been developed. For example, [54] builds on the basic ideas of Daws, incorporating various optimisations and integrating bisimulation minimisation to improve efficiency. This was implemented in the tool PARAM [52] and later also added to PRISM [80]. Since then, further improvements to parametric model checking of DTMCs have been proposed [65], including the use of strongly connected component decompositions and optimised approaches to the generation of rational functions; these have been implemented in the PROPhESY tool [37].

Below, we explain the key definitions and illustrate the approach on some examples. We refer the reader to the references above for more details.

Definition 3.27 (*Rational function*) Let $V = \{x_1, \ldots, x_n\}$ be a set of real-valued variables. A *rational function* f over V is a function of the form $f(x_1, \ldots, x_n) = g_1(x_1, \ldots, x_n)/g_2(x_1, \ldots, x_n)$ where g_1 and g_2 are polynomials each taking the form

$\sum_{i=1}^{m} a_i \, x_1^{k_{i,1}} \cdots x_n^{k_{i,n}}$ for $m \in \mathbb{N}$, $a_i \in \mathbb{R}$ for $1 \leqslant i \leqslant m$ and $k_{i,j} \in \mathbb{N}$ for $1 \leqslant i \leqslant m$ and $1 \leqslant j \leqslant n$. The set of all rational functions over variables V is denoted $\mathcal{F}_V$.

Given a rational function f over the variables V, a subset $V' \subseteq V$ of the variables and an evaluation $u : V' \rightarrow \mathbb{R}$ of V', we let $f[V'/u]$ denote the rational function obtained from f by substituting any occurrence of a variable $v' \in V'$ with the value $u(v')$. If $V' = V$, then u is a *total evaluation* and $f[V'/u]$ is a rational constant.

Definition 3.28 (*Parametric DTMC*) A *parametric DTMC* (PDTMC) is a tuple $\mathsf{D} = (S, \bar{s}, \mathbf{P}, L, V)$ where the set of states S, initial state $\bar{s}$ and labelling L are as for a DTMC (see Definition 3.1) and:

- $V = \{x_1, \ldots, x_n\}$ is a set of real-valued variables called *parameters*;
- $\mathbf{P} : S \times S \rightarrow \mathcal{F}_V$ is a probabilistic transition matrix mapping each pair of states to a rational function over the parameters.

A PDTMC retains the same basic structure as a DTMC, but transition probabilities are expressed as functions of its parameters. Evaluations for this set of parameters (satisfying certain conditions) then induce normal DTMCs.

Definition 3.29 (*Induced DTMC*) Let $\mathsf{D} = (S, \bar{s}, \mathbf{P}, L, V)$ be a PDTMC and $u : V \rightarrow \mathbb{R}$ be a total evaluation of its parameters. Let $\mathbf{P}_u(s, s) : S \times S \rightarrow \mathbb{R}$ be the matrix defined by $\mathbf{P}_u(s, s) = \mathbf{P}(s, s)[V/u]$. We say that the evaluation u is *well defined* for D if $\mathbf{P}_u(s, s') \in [0, 1]$ and $\sum_{s' \in S} \mathbf{P}_u(s, s') = 1$ for all $s, s' \in S$. In this case, the *induced DTMC* of the evaluation u is the DTMC $\mathsf{D}_u = (S, \bar{s}, \mathbf{P}_u, L)$.

Since the behaviour of a DTMC can be qualitatively different if its underlying transition graph changes, we assume that parameter evaluations u are *graph preserving*, meaning that, $\mathbf{P}(s, s') \neq 0$ implies $\mathbf{P}_u(s, s') > 0$ for all $s, s' \in S$. The basic property of interest for parametric DTMCs can then be defined as follows.

Definition 3.30 (*Probabilistic reachability for PDTMCs*) Let $\mathsf{D} = (S, \bar{s}, \mathbf{P}, L, V)$ be a PDTMC and $\mathsf{a} \in AP$ be an atomic proposition. The probabilistic reachability problem is to find a rational function $f \in \mathcal{F}_V$ such that, for any well-defined and graph preserving evaluation $u : V \rightarrow \mathbb{R}$ for D, we have:

$$f[V/u] = Pr_{\mathsf{D}_u, \bar{s}} \{ \pi \in IPaths_{\mathsf{D}_u}(\bar{s}) \mid \mathsf{D}_u, \pi \models \mathtt{F}\ \mathsf{a} \} .$$

Parametric probabilistic model checking of DTMCs has been applied to various problems, including model repair [16] and sensitivity analysis [43]. Below, we illustrate its usage on a simple example.

Example 10 We return to our running example, and adapt the DTMC version of the robot navigation model presented in Example 1 (see Fig. 3.1). Figure 3.13 (left) shows a modified version of this model, to which we have added to parameters p and q which occur in some of the transition probabilities. The original DTMC results from the parameter evaluation u that chooses $u(p) = 0.05$ and $u(q) = 0.75$.

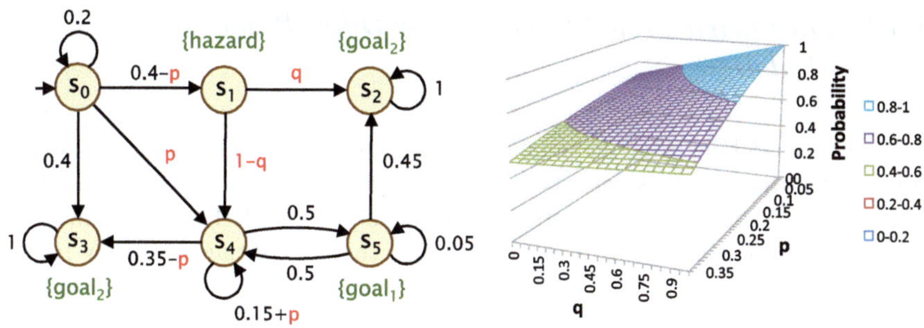

Fig. 3.13 Parametric model checking applied to an adapted version of the DTMC from Fig. 3.1

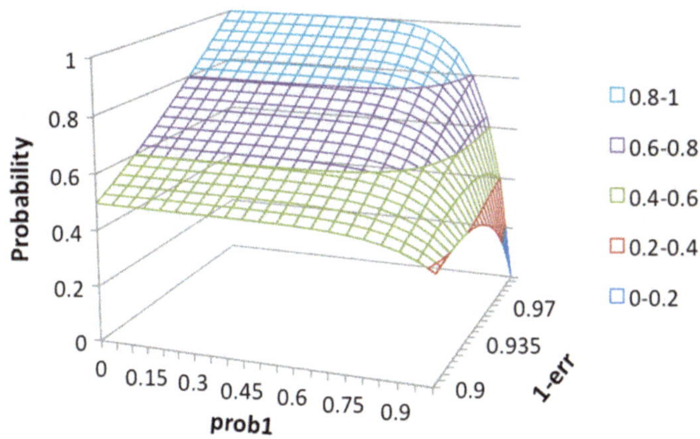

Fig. 3.14 Parametric model checking results for the NAND case study

We consider the property $\mathtt{P}_{=?}[\neg \mathsf{goal}_1 \ \mathtt{U} \ \mathsf{goal}_2]$, i.e. the probability of reaching goal 2 before goal 1. Applying parametric probabilistic model checking yields the rational function $(25 \cdot p \cdot q + 40 \cdot p - 10 \cdot q - 24)/(40 \cdot p - 34)$ as a result, which is plotted for the valid ranges of p and q in Fig. 3.13 (right). ■

Example 11 As a second example, we revisit the NAND multiplexing case study described in Sect. 3.2.1.3. There are two parameters we consider for this case study: *err* representing the probability that a NAND gate is unreliable and *prob1* the probability the initial input is correct (takes the value `true`). In Fig. 3.14, for the case when there are five copies of the inputs and outputs ($N = 5$) and one restorative stage ($M = 1$), we have plotted the probability that the error is less than 10 percent (the first property considered in Sect. 3.2.1.3) as the parameters *err* and *prob1* vary. ■

3.6.2 *Parametric Model Checking for Other Probabilistic Models*

Parametric model checking techniques have also been developed for several of the other probabilistic models described in this chapter. For example, Hahn et al. [54] extend the approach described above to the analysis of MDPs, where the non-deterministic choices are encoded as additional (binary) parameters. They found, however, that this method was limited by the number of non-deterministic choices available in a state and the fact that it could not be extended to nested properties.

They have since proposed an alternative method for parametric model checking of MDPs [51]. Instead of finding a rational function corresponding to an optimal probability or expected reward value, this approach finds parameter values for which a given property holds (or does not hold), i.e. it solves the parameter synthesis problem. This is achieved by repeatedly dividing the parameter space into regions (hyper-rectangles) until regions are found over which the property of interest holds or does not hold. Checking this requirement over a region is performed by first finding an optimal strategy for the 'middle' point of the region, using standard (non-parametric) MDP model checking of MDPs, and then performing parametric model checking on the induced (parametric) DTMC of this strategy over the region.

Techniques also exist for CTMCs. Parametric model checking of unbounded properties for CTMCs can use the same methods as those developed for DTMCs. For time-bounded properties, [59] proposes an approach which approximates the set of parameter values for which a time-bounded probabilistic reachability property holds, based on a discretisation of the parameter space. We also mention [24, 26], which allows for precise parametric model checking of time bounded properties of CTMCs. This works by iteratively dividing the parameter space into regions through the computation of upper and lower bound approximations for the time-bounded reachability probability of interest over the regions.

Finally, concerning PTAs, both [11] and [69] study the problem of synthesising timing constraints of a PTA to ensure the satisfaction of a given property. The approach of [11] is based on the inverse method for parametric (non-probabilistic) timed automata [10], while [69] extends the forwards reachability [84] and game-based [79] approaches for model checking PTAs.

3.7 Future Challenges and Directions

This chapter has provided an overview of probabilistic model checking and surveyed some of the significant advances that have been made in the area in recent years. Probabilistic model checking has shown itself to be a powerful, flexible and broadly applicable verification technique, but a number of key challenges remain and work continues on many fronts to improve the state of the art.

As with most areas of formal verification, a recurring limitation of probabilistic model checking is its scalability to large, complex systems. We have discussed various efforts to tackle this problem in earlier sections. Another related and fundamental issue, which is true of any model-based analysis technique, is that the results of verification are only as reliable as the model itself. For models with quantitative aspects such as probability and time, which may be difficult to measure accurately, this is particularly pertinent.

We conclude this chapter by highlighting some of the key challenges and research directions in the area of probabilistic model checking, many of which aim to tackle these issues.

Hybrid systems. Probabilistic model checking has many applications in the domain of embedded and cyber-physical systems, for example in the verification of sensor networks or robotic applications. In this setting, the interaction of (discrete) computerised systems with their (continuous) environment becomes a crucial issue. Such hybrid systems (or cyber-physical systems) raise new challenges because they require more powerful models such as stochastic hybrid automata.

Hybrid automata allow both discrete behaviour and continuous flows defined through differential equations, for example to model thermodynamics. The verification of hybrid automata is in general undecidable, therefore the analysis is restricted to certain subclasses and considering only approximate results. Early work on probabilistic hybrid automata concerned decidability results for different subclasses [102]. Recent work [56, 113] combines abstraction approaches for non-probabilistic hybrid automata [4, 98] with the abstraction-refinement approaches for MDPs [34, 75] discussed in Sect. 3.4. We also mention [38], where two approximation techniques for classical hybrid automata are extended to the probabilistic case and [47] which, using stochastic satisfiability modulo theories, presents a decision procedure for verifying time-bounded properties.

Probabilistic software and programs. Although the modelling languages of tools such as PRISM are sufficiently expressive for many purposes, direct support for the probabilistic model checking of mainstream programming languages such as C or Java or of system-level modelling languages such as SystemC will be required for the verification of real applications. Programs in these languages yield extremely large, or infinite state, models, which need dedicated techniques to tackle. A related area, which has attracted interest in recent years, is the verification of probabilistic programming languages [71], which have applications both for the specification of randomised or probabilistic software and for the development of probabilistic models used for inference and machine learning.

Ubiquitous computing. The vision of ubiquitous or pervasive computing sees thousands of computerised devices integrating seamlessly in daily life. This emphasises the need for techniques to ensure their correctness, but also demands the development of new modelling formalisms and analysis techniques that can handle both the dynamic nature and the enormous scale of these systems. One key aspect to modelling ubiquitous computing devices is *autonomous behaviour*, as can be seen in, for example, driverless cars and drone missions. In addition, we need to model the *constrained resources* (often devices have limited memory and CPU processing and are battery powered) and the fact that devices need to be *adaptive* as requirements and the environment evolve.

Partial observability. In this chapter, we have assumed that the state of the system and history are fully visible to a strategy when making decisions. However, in many situations, this is unrealistic, for example, to verify that a security protocol is functioning correctly, it may be essential to model the fact that some data held by a participant is not externally visible, or, when synthesising a controller for a robot, the controller may not be implementable in practice if it bases its decisions on information that cannot be physically observed.

Partially observable MDPs (POMDPs) are a natural extension of MDPs for modelling such strategies and they are widely used in areas such as planning and artificial intelligence, but verification of POMDPs is considerably more difficult than MDPs since key problems are undecidable [87]. Work in this area towards practical verification of POMDPs includes [27, 93, 105].

Robustness and uncertainty. In many potential applications, such as the generation of controllers in embedded systems, it may be difficult to formulate a precise model of the stochastic behaviour of the system's environment. Thus, developing appropriate models of uncertainty, and corresponding methods to synthesise strategies that are robust in these environments, is important. Developing more sophisticated approaches is an active area of research [96, 112].

Counterexamples. One final challenge is to improve the quality and usefulness of the results that are generated by probabilistic model checking. One of the main reasons for the success of non-probabilistic model checking is the generation of counterexamples which provide, when the property being verified does not hold, evidence of this violation. This evidence is usually in the form of a path demonstrating the violation. In the probabilistic case, there is the complication that, to refute a property, a single path is in general not sufficient as more than one path can contribute to the probability of the property not holding. Initial research [58], focused on DTMCs and reachability properties and generating a finite set of paths. Recent research has focused on generating a more useful representation for counterexamples, including regular expressions, hierarchical representations and critical sub-systems, for further information see, for example, the survey [1].

Acknowledgements This work was supported by the ERC Advanced Investigators Grant VERIWARE, the EPSRC Mobile Autonomy Programme Grant EP/M019918/1, the EU FP7-funded project HIERATIC and the DARPA-funded BRASS project.

References

1. E. Ábrahám, B. Becker, C. Dehnert, N. Jansen, J.-P. Katoen, R. Wimmer, Counterexample generation for discrete-time Markov models: an introductory survey, in *Formal Methods for the Design of Computer, Communication, and Software Systems (SFM'14)*, ed. By M. Bernardo, F. Damiani, R. Haehnle, E. Johnsen, I. Schaefer. LNCS, vol. 8483 (Springer, 2014), pp. 65–121
2. R. Alur, C. Courcoubetis, D. Dill, Model-checking for probabilistic real-time systems, in *Proceedings of the 19th International Colloq Automata, Languages and Programming (ICALP'91)*. LNCS, vol. 510, (Springer, 1991), pp. 115–136
3. R. Alur, C. Courcoubetis, D. Dill, Model checking in dense real time. Inf. Comput. **104**(1), 2–34 (1993)
4. R. Alur, T. Dang, F. Ivancic, Predicate abstraction for reachability analysis of hybrid systems. ACM Trans. Embed. Comput. Syst. **5**(1), 152–199 (2006)
5. R. Alur, D. Dill, A theory of timed automata. Theor. Comput. Sci. **126**, 183–235 (1994)
6. R. Alur, T. Henzinger, O. Kupferman, Alternating-time temporal logic. J. ACM **49**(5), 672–713 (2002)
7. R. Alur, T. Henzinger, S. Rajamani, Symbolic exploration of transition hierarchies, in *Proceedings of the 4th International Conference Tools and Algorithms for the Construction and Analysis of Systems (TACAS'98)*. LNCS, vol. 1384, (Springer, 1998), pp. 330–344
8. R. Alur, S. La Torre, G. Pappas, Optimal paths in weighted timed automata. Theor. Comput. Sci. **318**(3), 297–322 (2004)
9. R. Alur, A. Trivedi, Relating average and discounted costs for quantitative analysis of timed systems, in *Proceedings of the 11th International Conference Embedded Software (EMSOFT'11)* (ACM, 2011), pp. 165–174
10. E. André, T. Chatain, E. Encrenaz, L. Fribourg, An inverse method for parametric timed automata. Int. J. Found. Comput. Sci. **20**(5), 819–836 (2009)
11. E. André, L. Fribourg, J. Sproston, An extension of the inverse method to probabilistic timed automata. Form. Methods Syst. Des. **42**(2), 119–145 (2013)
12. C. Baier, E. Clarke, V. Hartonas-Garmhausen, M. Kwiatkowska, M. Ryan, Symbolic model checking for probabilistic processes, in *Proceedings of the 24th International Colloquium Automata, Languages and Programming (ICALP'97)*, ed. By P. Degano, R. Gorrieri, A. Marchetti-Spaccamela. LNCS, vol. 1256 (Springer, 1997), pp. 430–440
13. C. Baier, M. Größer, M. Leucker, B. Bollig, F. Ciesinski, Controller synthesis for probabilistic systems, in *Proceedings of the 3rd IFIP International Conference Theoretical Computer Science (TCS'06)*, ed. By J.-J. Lévy, E. Mayr, J. Mitchell (Kluwer, 2004), pp. 493–5062
14. C. Baier, B. Haverkort, H. Hermanns, J.-P. Katoen, Model-checking algorithms for continuous-time Markov chains. IEEE Trans. Softw. Eng. **29**(6), 524–541 (2003)
15. C. Baier, J.-P. Katoen, *Principles of Model Checking* (MIT Press, Cambridge, 2008)
16. E. Bartocci, R. Grosu, P. Katsaros, C. Ramakrishnan, S. Smolka, Model repair for probabilistic systems, in *Proceedings of the 17th International Conference Tools and Algorithms for the Construction and Analysis of Systems (TACAS'11)*, ed. By P. Abdulla, K. Leino. LNCS, vol. 6605 (Springer, 2011), pp. 326–340
17. D. Beauquier, Probabilistic timed automata. Theor. Comput. Sci. **292**(1), 65–84 (2003)
18. G. Behrmann, A. Cougnard, A. David, E. Fleury, K. Larsen, D. Lime, UPPAAL-Tiga: time for playing games!, in *Proceedings of the 19th International Conference Computer Aided Verification (CAV'07)*. LNCS, vol. 4590 (Springer, 2007), pp. 121–125
19. G. Behrmann, A. Fehnker, T. Hune, K. Larsen, P. Pettersson, J. Romijn, Efficient guiding towards cost-optimality in UPPAAL, in *Proceedings of the 7th International Conference Tools and Algorithms for the Construction and Analysis of Systems (TACAS'01)*, ed. By T. Margaria, W. Yi. LNCS, vol. 2031 (Springer, 2001), pp. 174–188
20. R. Bellman, *Dynamic Programming* (Princeton University Press, New Jersey, 1957)
21. P. Billingsley, *Probability and Measure* (Wiley, New Jersey, 1995)
22. P. Bouyer, U. Fahrenberg, K. Larsen, N. Markey, Quantitative analysis of real-time systems using priced timed automata. Commu. ACM **54**(9), 78–87 (2011)

23. T. Brázdil, V. Brožek, V. Forejt, A. Kučera, Stochastic games with branching-time winning objectives, in *Proceedings of the 21th IEEE Symposium Logic in Computer Science (LICS'06)* (IEEE Computer Society, 2006), pp. 349–358
24. L. Brim, M. Češka, D.V.S. Dražan, Exploring parameter space of stochastic biochemical systems using quantitative model checking, in *Proceedings of the 25th International Conference Computer Aided Verification (CAV'13)*. LNCS, vol. 8044 (Springer, 2013), pp. 107–123
25. P. Buchholz, E.M. Hahn, H. Hermanns, L. Zhang, Model checking algorithms for CTMDPs, in *Proceedings of the 23rd International Conference Computer Aided Verification (CAV'11)*, ed. By G. Gopalakrishnan, S. Qadeer. LNCS, vol. 6806 (Springer, 2011), pp. 225–242
26. M. Češka, F. Dannenberg, M. Kwiatkowska, N. Paoletti, Precise parameter synthesis for stochastic biochemical systems, in *Proceedings of the 12th International Conference Computational Methods in Systems Biology (CMSB'14)*, ed. By P. Mendes, J. Dada, K. Smallbone. LNCS/LNBI, vol. 8859 (Springer, 2014), pp. 86–98
27. K. Chatterjee, M. Chmelík, R. Gupta, A. Kanodia, Qualitative analysis of POMDPs with temporal logic specifications for robotics applications, in *Proceedings of the IEEE International Conference Robotics and Automation, (ICRA'15)* (IEEE Computer Society, 2015), pp. 325–330
28. T. Chen, V. Forejt, M. Kwiatkowska, D. Parker, A. Simaitis, Automatic verification of competitive stochastic systems. Form. Methods Syst. Des. **43**(1), 61–92 (2013)
29. S. Cheshire, B. Adoba, E. Gutterman, Dynamic configuration of IPv4 link local addresses. http://www.ietf.org/rfc/rfc3927.txtwww.ietf.org/rfc/rfc3927.txt
30. E. Clarke, O. Grumberg, S. Jha, Y. Lu, H. Veith, Counterexample-guided abstraction refinement, in *Proceedings of the 12th International Conference Computer Aided Verification (CAV'00)*, ed. By A. Emerson, A. Sistla. LNCS, vol. 1855 (Springer, 2000), pp. 154–169
31. A. Condon, The complexity of stochastic games. Inf. Comput. **96**(2), 203–224 (1992)
32. A. Condon, On algorithms for simple stochastic games, *Advances in computational complexity theory*. DIMACS Series in Discrete Mathematics and Theoretical Computer Science **13**, 51–73 (1993)
33. M. Daniele, F. Giunchiglia, M. Vardi, Improved automata generation for linear temporal logic, in *Proceedings of the 11th International Conference Computer Aided Verification (CAV'99)*, ed. By N. Halbwachs, D. Peled. LNCS, vol. 1633 (Springer, 1999), pp. 249–260
34. P. D'Argenio, B. Jeannet, H. Jensen, K. Larsen, in *Reachability analysis of probabilistic systems by successive refinements, Proceedings of the 1st Joint International Workshop Process Algebra and Probabilistic Methods, Performance Modelling and Verification (PAPM/PROBMIV'01)*, ed. By L. de Alfaro, S. Gilmore. LNCS, vol. 2165 (Springer, 2001), pp. 39–56
35. C. Daws, Symbolic and parametric model checking of discrete-time Markov chains, in *Proceedings of the 1st International Colloquium Theoretical Aspects of Computing (ICTAC'04)*, ed. By Z. Liu, K. Araki. LNCS, vol. 3407 (Springer, 2004), pp. 280–294
36. L. de Alfaro, Formal Verification of Probabilistic Systems. Ph.D. thesis, Stanford University, 1997
37. C. Dehnert, S. Junges, N. Jansen, F. Corzilius, M. Volk, H. Bruintjes, J.-P. Katoen, E. Ábrahám, PROPhESY: a PRObabilistic ParamEter SYnthesis tool, in *Proceedings of the 27th International Conference Computer Aided Verification (CAV'15)*. LNCS, vol. 9206 (Springer, 2015), pp. 214–231
38. J. Desharnais, J. Assouramou, Analysis of non-linear probabilistic hybrid systems, in *Proceedings of the 9th Workshop Quantitative Aspects of Programming Languages (QAPL'11)*. EPTCS, vol. 57 (2011), pp. 104–119
39. A. Donaldson, A. Miller, Symmetry reduction for probabilistic model checking using generic representatives, in *Proceedings of the 4th International Symposium Automated Technology for Verification and Analysis (ATVA'06)*, ed. By S. Graf, W. Zhang. LNCS, vol. 4218 (Springer, 2006), pp. 9–23
40. S. Donatelli, S. Haddad, J. Sproston, Model checking timed and stochastic properties with CSL^{ta}. IEEE Trans. Softw. Eng. **35**(2), 224–240 (2008)

41. C. Eisentraut, H. Hermanns, L. Zhang, On probabilistic automata in continuous time, in *Proceedings of the 25th Annual IEEE Symposium Logic in Computer Science (LICS'10)* (IEEE Computer Society, 2010), pp. 342–351
42. K. Etessami, M. Kwiatkowska, M. Vardi, M. Yannakakis, Multi-objective model checking of Markov decision processes. Logical Methods Comput. Sci. **4**(4), 1–21 (2008)
43. A. Filieri, G. Tamburrelli, C. Ghezzi, Supporting self-adaptation via quantitative verification and sensitivity analysis at run time. IEEE Trans. Softw. Eng. **42**(1), 75–99 (2016)
44. V. Forejt, M. Kwiatkowska, G. Norman, D. Parker, Automated verification techniques for probabilistic systems, in *Formal Methods for Eternal Networked Software Systems (SFM'11)*, ed. By M. Bernardo, V. Issarny. LNCS, vol. 6659 (Springer, 2011), pp. 53–113
45. V. Forejt, M. Kwiatkowska, G. Norman, D. Parker, H. Qu, Quantitative multi-objective verification for probabilistic systems, in *Proceedings of the 17th International Conference Tools and Algorithms for the Construction and Analysis of Systems (TACAS'11)*, ed. By P. Abdulla, K. Leino. LNCS, vol. 6605 (Springer, 2011), pp. 112–127
46. V. Forejt, M. Kwiatkowska, D. Parker, Pareto curves for probabilistic model checking, in *Proceedings of the 10th International Symposium Automated Technology for Verification and Analysis (ATVA'12)*, ed. By S. Chakraborty, M. Mukund. LNCS, vol. 7561 (Springer, 2012), pp. 317–332
47. M. Fränzle, T. Teige, A. Eggers, Engineering constraint solvers for automatic analysis of probabilistic hybrid automata. J. Logic Algebr. Progr. **79**(7), 436–466 (2010)
48. S. Graf, H. Saidi, Construction of abstract state graphs with PVS, in *Proceedings of the 9th International Conference Computer Aided Verification (CAV'97)*, ed. By O. Grumberg. LNCS, vol. 1254 (Springer, 1997), pp. 72–83
49. D. Gross, D. Miller, The randomization technique as a modeling tool and solution procedure for transient Markov processes. Oper. Res. **32**(2), 343–361 (1984)
50. M. Größer, C. Baier, Partial order reduction for Markov decision processes: a survey, in *Proceedings of the 4th International Symposium Formal Methods for Component and Objects (FMCO'05)*, ed. By F. de Boer, M. Bonsangue, S. Graf, W.-P. de Roever. LNCS, vol. 4111 (Springer, 2006), pp. 408–427
51. E.M. Hahn, T. Han, L. Zhang, Synthesis for PCTL in parametric Markov decision processes, in *Proceedings of the 3rd NASA Formal Methods Symposium (NFM'11)*. LNCS, vol. 6617 (Springer, 2011)
52. E.M. Hahn, H. Hermanns, B. Wachter, L. Zhang, PARAM: a model checker for parametric Markov models, in *Proceedings of the 22nd International Conference Computer Aided Verification (CAV'10)*. LNCS, vol. 6174 (Springer, 2010), pp. 660–664
53. E.M. Hahn, H. Hermanns, B. Wachter, L. Zhang, PASS: abstraction refinement for infinite probabilistic models, in *Proceedings of the 16th International Conference Tools and Algorithms for the Construction and Analysis of Systems (TACAS'10)*, ed. By J. Esparza, R. Majumdar. LNCS, vol. 6105 (Springer, 2010), pp. 353–357
54. E.M. Hahn, H. Hermanns, L. Zhang, Probabilistic reachability for parametric Markov models. Int. J. Softw. Tools Technol. Trans. (STTT) **13**(1), 3–19 (2011)
55. E.M. Hahn, Y. Li, S. Schewe, A. Turrini, L. Zhang, iscasMc: a web-based probabilistic model checker, in *Proceedings of the 19th International Symposium on Formal Methods (FM'14)* (2014), pp. 312–317
56. E.M. Hahn, G. Norman, D. Parker, B. Wachter, L. Zhang, Game-based abstraction and controller synthesis for probabilistic hybrid systems, in *Proceedings of the 8th International Conference Quantitative Evaluation of SysTems (QEST'11)* (IEEE Computer Society Press, 2011), pp. 69–78
57. J. Han, P. Jonker, A system architecture solution for unreliable nanoelectronic devices. IEEE Trans. Nanotechnol. **1**, 201–208 (2002)
58. T. Han, J.-P. Katoen, B. Damman, Counterexample generation in probabilistic model checking. IEEE Trans. Softw. Eng. **35**(2), 241–257 (2009)
59. T. Han, J.-P. Katoen, A. Mereacre, Approximate parameter synthesis for probabilistic time-bounded reachability, in *Proceedings of the IEEE Real-Time Systems Symposium (RTSS 08)* (IEEE Computer Society Press, 2008), pp. 173–182

60. H. Hansson, B. Jonsson, A logic for reasoning about time and reliability. Form. Asp. Comput. **6**(5), 512–535 (1994)
61. A. Hartmanns, H. Hermanns, A modest approach to checking probabilistic timed automata, in *Proceedings of the 6th International Conference Quantitative Evaluation of Systems (QEST'09)* (2009). To appear
62. H. Hermanns, *Interactive Markov Chains and the Quest for Quantified Quality*. LNCS, vol. 2428 (Springer, New York, 2002)
63. H. Hermanns, B. Wachter, L. Zhang, Probabilistic CEGAR, in *Proceedings of the 20th International Conference Computer Aided Verification (CAV'08)*, ed. By A. Gupta, S. Malik. LNCS, vol. 5123 (Springer, 2008), pp. 162–175
64. R. Howard, *Dynamic Programming and Markov Processes* (The MIT Press, Cambridge, 1960)
65. N. Jansen, F. Corzilius, M. Volk, R. Wimmer, E. Ábrahám, J.-P. Katoen, B. Becker, Accelerating parametric probabilistic verification, in *Proceedings of the 11th International Conference Quantitative Evaluation of Systems (QEST'14)* (2014), pp. 404–420
66. B. Jeannet, P. D'Argenio, K. Larsen, Rapture: a tool for verifying Markov decision processes, in *Proceedings of the Tools Day, affiliated to 13th International Conference Concurrency Theory (CONCUR'02)*, ed. By I. Cerna. Technical Report FIMU-RS-2002-05, Faculty of Informatics Masaryk University (2002), pp. 84–98
67. A. Jensen, Markoff chains as an aid in the study of Markoff processes. Skandinavisk Aktuarietidskrift **36**, 87–91 (1953)
68. H. Jensen, Model checking probabilistic real time systems, in *Proceedings of the 7th Nordic Workshop Programming Theory* (1996), pp. 247–261
69. A. Jovanovic, M. Kwiatkowska, Parameter synthesis for probabilistic timed automata using stochastic games, in *Proceedings of the 8th International Workshop Reachability Problems (RP'14)*, ed. By J. Ouaknine, I. Potapov, J. Worrell. LNCS, vol. 8762, (Springer, 2014), pp. 176–189
70. M. Jurdziński, M. Kwiatkowska, G. Norman, A. Trivedi, Concavely-priced probabilistic timed automata, in *Proceedings of the 20th International Conference Concurrency Theory (CONCUR'09)*, ed. By M. Bravetti, G. Zavattaro. LNCS, vol. 5710 (Springer, 2009), pp. 415–430
71. J.-P. Katoen, Probabilistic programming: a true challenge in verification, in *Proceedings of the 13th International Symposium on Automated Technology for Verification and Analysis (ATVA'15)*. LNCS (Springer, 2015), pp. 1–3
72. J.-P. Katoen, T. Kemna, I. Zapreev, D. Jansen, Bisimulation minimisation mostly speeds up probabilistic model checking, in *Proceedings of the 13th International Conference Tools and Algorithms for the Construction and Analysis of Systems (TACAS'07)*, ed. By O. Grumberg, M. Huth. LNCS, vol. 4424 (Springer, 2007), pp. 87–101
73. J.-P. Katoen, I. Zapreev, E.M. Hahn, H. Hermanns, D. Jansen, The ins and outs of the probabilistic model checker MRMC, in *Proceedings of the 6th International Conference Quantitative Evaluation of Systems (QEST'09)* (IEEE Computer Society Press, 2009), pp. 167–176
74. M. Kattenbelt, M. Kwiatkowska, G. Norman, D. Parker, Abstraction refinement for probabilistic software, in *Proceedings of the 10th International Conference Verification, Model Checking, and Abstract Interpretation (VMCAI'09)*, ed. By N. Jones, M. Muller-Olm. LNCS, vol. 5403 (Springer, 2009), pp. 182–197
75. M. Kattenbelt, M. Kwiatkowska, G. Norman, D. Parker, A game-based abstraction-refinement framework for Markov decision processes. Form. Methods Syst. Des. **36**(3), 246–280 (2010)
76. J. Kemeny, J. Snell, A. Knapp, *Denumerable Markov Chains*, 2nd edn. (Springer, Heidelberg, 1976)
77. M. Kwiatkowska, G. Norman, D. Parker, Symmetry reduction for probabilistic model checking, in *Proceedings of the 18th International Conference Computer Aided Verification (CAV'06)*, ed. By T. Ball, R. Jones. LNCS, vol. 4114 (Springer, 2006), pp. 234–248
78. M. Kwiatkowska, G. Norman, D. Parker, Stochastic model checking, in *Formal Methods for the Design of Computer, Communication and Software Systems: Performance Evaluation (SFM'07)*, ed. By M. Bernardo, J. Hillston. LNCS (Tutorial Volume), vol. 4486 (Springer, 2007), pp. 220–270

79. M. Kwiatkowska, G. Norman, D. Parker, Stochastic games for verification of probabilistic timed automata, in *Proceedings of the 7th International Conference Formal Modelling and Analysis of Timed Systems (FORMATS'09)*, ed. By J. Ouaknine, F. Vaandrager. LNCS, vol. 5813 (Springer, 2009), pp. 212–227
80. M. Kwiatkowska, G. Norman, D. Parker, PRISM 4.0: verification of probabilistic real-time systems, in *Proceedings of the 23rd International Conference Computer Aided Verification (CAV'11)*, ed. By G. Gopalakrishnan, S. Qadeer. LNCS, vol. 6806 (Springer, 2011), pp. 585–591
81. M. Kwiatkowska, G. Norman, D. Parker, H. Qu, Compositional probabilistic verification through multi-objective model checking. Inf. Comput. **232**, 38–65 (2013)
82. M. Kwiatkowska, G. Norman, D. Parker, J. Sproston, Performance analysis of probabilistic timed automata using digital clocks. Form. Methods Syst. Des. **29**, 33–78 (2006)
83. M. Kwiatkowska, G. Norman, R. Segala, J. Sproston, Verifying quantitative properties of continuous probabilistic timed automata, in *In Proceedings of the 11th International Conference Concurrency Theory (CONCUR'00)*, ed. By C. Palamidessi. LNCS, vol. 1877 (Springer, 2000), pp. 123–137
84. M. Kwiatkowska, G. Norman, R. Segala, J. Sproston, Automatic verification of real-time systems with discrete probability distributions. Theor. Comput. Sci. **282**, 101–150 (2002)
85. M. Kwiatkowska, G. Norman, J. Sproston, F. Wang, Symbolic model checking for probabilistic timed automata. Inf. Comput. **205**(7), 1027–1077 (2007)
86. M. Kwiatkowska, D. Parker, C. Wiltsche, PRISM-games 2.0: a tool for multi-objective strategy synthesis for stochastic games, in *Proceedings of the 22nd International Conference Tools and Algorithms for the Construction and Analysis of Systems (TACAS'16)*. LNCS (Springer, 2016)
87. O. Madani, S. Hanks, A. Condon, On the undecidability of probabilistic planning and related stochastic optimization problems. Artif. Intell. **147**(1–2), 5–34 (2003)
88. O. Maler, K. Larsen, B. Krogh, On zone-based analysis of duration probabilistic automata, in *Proceedings of the 12th International Workshop Verification of Infinite-State Systems (INFINITY'10)*. EPTCS, vol. 39 (2010), pp. 33–46
89. R. Milner, Calculi for synchrony and asynchrony. Theor. Comput. Sci. **25**(3), 267–310 (1993)
90. R. Nicola, F. Vaandrager, Action versus state based logics for transition systems, in *Proceedings of the LITP Spring School on Theoretical Computer Science: Semantics of Systems of Concurrent Processes*, ed. By I. Guessarian (Springer, 1990), pp. 407–419
91. G. Norman, D. Parker, M. Kwiatkowska, S. Shukla, Evaluating the reliability of NAND multiplexing with PRISM. IEEE Trans. Comput. Aided Des. Integr. Circuits Syst. **24**(10), 1629–1637 (2005)
92. G. Norman, D. Parker, J. Sproston, Model checking for probabilistic timed automata. Form. Methods Syst. Des. **43**(2), 164–190 (2013)
93. G. Norman, D. Parker, X. Zou, Verification and control of partially observable probabilistic real-time systems, in *Proceedings of the 13th International Conference Formal Modelling and Analysis of Timed Systems (FORMATS'15)*, ed. By S. Sankaranarayanan, E. Vicario. LNCS, vol. 9268 (Springer, 2015), pp. 240–255
94. D. Parker, Implementation of Symbolic Model Checking for Probabilistic Systems. Ph.D. thesis, University of Birmingham, 2002
95. A. Pnueli, The temporal semantics of concurrent programs. Theor. Comput. Sci. **13**, 45–60 (1981)
96. A. Puggelli, W. Li, A. Sangiovanni-Vincentelli, S. Seshia, Polynomial-time verification of PCTL properties of MDPs with convex uncertainties, in *Proceedings of the 25th International Conference Computer Aided Verification (CAV'13)*. LNCS, vol. 8044 (Springer, 2013), pp. 527–542
97. M. Puterman, *Markov Decision Processes: Discrete Stochastic Dynamic Programming* (Wiley, New Jersey, 1994)
98. S. Ratschan, Z. She, Safety verification of hybrid systems by constraint propagation-based abstraction refinement. ACM Trans. Embed. Comput. Syst. **6**(1) (2007)

99. A.W. Roscoe, *The Theory and Practice of Concurrency* (Prentice-Hall, New Jersey, 1997)
100. R. Segala, Modelling and verification of randomized distributed real time systems. Ph.D. thesis, Massachusetts Institute of Technology, 1995
101. R. Segala, N. Lynch, Probabilistic simulations for probabilistic processes. Nordic J. Comput. **2**(2), 250–273 (1995)
102. J. Sproston, Decidable model checking of probabilistic hybrid automata, in *Proceedings of the International Symposium on Formal Techniques in Real-Time and Fault Tolerant Systems (FTRTFT'00)*, ed. By M. Joseph. LNCS, vol. 1926 (Springer, 2000), pp. 31–45
103. J. Sun, Y. Liu, J.S. Dong, J. Pang, Pat: towards flexible verification under fairness, in *Proceedings of the 21st International Conference Computer Aided Verification (CAV'09)*. LNCS, vol. 5643 (Springer, 2009), pp. 709–714
104. M. Svorenova, M. Kwiatkowska, Quantitative verification and strategy synthesis for stochastic games. Eur. J. Control **30**, 15–30 (2016)
105. M. Svoreňová, M. Chmelík, K. Leahy, H. Eniser, K. Chatterjee, I. Černá, C. Belta, Temporal logic motion planning using POMDPs with parity objectives: case study paper, in *Proceedings of the 18th International Conference Hybrid Systems: Computation and Control (HSCC'15)* (ACM, 2015), pp. 233–238
106. S. Tripakis, The analysis of timed systems in practice. Ph.D. thesis, Université Joseph Fourier, Grenoble, 1998
107. S. Tripakis, S. Yovine, A. Bouajjan, Checking timed Buchi automata emptiness efficiently. Form. Methods Syst. Des. **26**(3), 267–292 (2005)
108. M. Vardi, P. Wolper, Reasoning about infinite computations. Inf. Comput. **115**(1), 1–37 (1994)
109. J. von Neumann, Probabilistic logics and synthesis of reliable organisms from unreliable components, in *Automata Studies*, ed. By C. Shannon, J. McCarthy (Princeton University Press, 1956), pp. 43–98
110. B. Wachter, L. Zhang, H. Hermanns, Probabilistic model checking modulo theories, in *Proceedings of the 4th International Conference Quantitative Evaluation of Systems (QEST'07)* (IEEE Computer Society Press, 2007), pp. 129–140
111. C. Wiltsche, Assume-Guarantee Strategy Synthesis for Stochastic Games. Ph.D thesis, University of Oxford, 2015
112. E. Wolff, U. Topcu, R. Murray, Robust control of uncertain Markov decision processes with temporal logic specifications, in *Proceedings of the IEEE 51st Annual Conference Decision and Control (CDC'12)* (Computer Society Press, 2012), pp. 3372–3379
113. L. Zhang, Z. She, S. Ratschan, H. Hermanns, E.M. Hahn, Safety verification for probabilistic hybrid systems. Eur. J. Control **18**(6), 572–587 (2012)
114. http://www.prismmodelchecker.org
115. http://www.prismmodelchecker.org/files/fsv-pmc/
116. http://www.prismmodelchecker.org/games
117. http://www.prismmodelchecker.org/other-tools.php

Author Biographies

Marta Kwiatkowska obtained her BSc/MSc degree in computer science from the Jagiellonian University in Cracow, Poland, in 1980, and PhD in computer science from the University of Leicester, UK, in 1989. She was an Assistant Professor at the Jagiellonian University from 1980 until 1988, Lecturer at the University of Leicester from 1986 until 1994, and Lecturer, Reader and Professor in the School of Computer Science at the University of Birmingham, UK, from 1994 to 2007. Since July 2007, she has been Professor of Computing Systems in the Department of Computer Science and Fellow of Trinity College, University of Oxford. She serves as Head of the Automated Verification Group and Deputy Head of Research. Her research interests focus on probabilistic modelling, verification and synthesis techniques, with applications in distributed systems, robotics, nanotechnology and biology.

Gethin Norman is a Lecturer in Computing Science at the University of Glasgow and was previously a senior post-doctoral researcher at the University of Oxford. He obtained a degree in mathematics from the University of Oxford in 1994 and a PhD in computer science from the University of Birmingham in 1997. The focus of his research has been on the theoretical underpinning of quantitative formal methods, particularly models and algorithms for real-time and probability, and quality of service properties. He is a key contributor to the probabilistic verification tool PRISM, developing many of PRISM's modelling case studies across a wide range of application domains.

David Parker completed a PhD in Computer Science at the University of Birmingham in 2003 and then worked as a post-doctoral researcher, first at Birmingham and then at the University of Oxford, between 2007 and 2012. He is currently a Senior Lecturer in Computer Science at the University of Birmingham. His main research interests are in the area of formal verification, with a particular focus on the analysis of quantitative aspects such as probabilistic and real-time behaviour, and he has published over 100 papers in this area. He also leads the development of the widely used probabilistic verification tool PRISM.

Chapter 4
Software in a Hardware View

New Models for HW-dependent Software in SoC Verification

Carlos Villarraga, Dominik Stoffel and Wolfgang Kunz

Abstract In current practices of SoC design a trend can be observed to integrate more and more low-level software components into the hardware at different levels of granularity. The implementation of important control functions is frequently shifted from the SoC's hardware into its firmware. This calls for new methods for verification based on a joint analysis of hardware and software. While most techniques of software verification operate at a hardware-independent level, this chapter elaborates on the possible merits of a hardware-dependent software view. The chapter reviews a recently developed model for formal verification of low-level embedded system software called *program netlist* and details on its applications. In particular, applications for speed-independent and cycle-accurate hardware/software integration are reported. For each studied scenario, this chapter describes how the different challenges of modeling the hardware/software interface can be solved by exploiting the characteristics of the program netlist. For speed-independent hardware/software interaction the equivalence checking problem is studied and results of our proposed solution are presented. For the case of a cycle-accurate hardware/software integration, a model for hardware/software co-verification is developed and experimentally evaluated by applying it to property checking.

4.1 Introduction

In recent years, the programmability of Systems-on-Chip (SoC) has increased continuously. This does not only allow for creating application software with growing complexity but also changes practices at lower design levels. In particular, a trend

C. Villarraga (✉) · D. Stoffel · W. Kunz
Department of Electrical and Computer Engineering, University of Kaiserslautern, Kaiserslautern, Germany
e-mail: villarraga@eit.uni-kl.de

D. Stoffel
e-mail: stoffel@eit.uni-kl.de

W. Kunz
e-mail: kunz@eit.uni-kl.de

R. Drechsler (ed.), *Formal System Verification*, DOI 10.1007/978-3-319-57685-5_4

towards a firmware-based design style can be observed. Certain control functions of an SoC module are no longer implemented in hardware but as firmware running on processors instantiated particularly for this purpose. Similarly, the implementation of SoC communication structures is shifted more and more from hardware to the low-level software of the system.

Tight coupling of hardware and software at a low level of granularity is also common when implementing the non-mainline functions of an SoC such as system reset, power management and the control of infrastructures for test and diagnosis. Therefore, also the verification of non-mainline functionality has become a concern in industry [11, 13].

This has created new interest in techniques for formal software verification, not only among software developers but also in the hardware design community. Traditional techniques for formal software verification, however, usually adopt a hardware-independent view when verifying software programs written in high-level languages such as C. This is reasonable for a wide range of applications where the main objective is to identify bugs that are specific to the software development process. However, in embedded system design, as a result of the trends described above, it is important to analyze the mutual effects of hardware and software on each other. Therefore, a hardware-dependent software view is needed where the behavior of a program is precisely described in terms of its effect on the underlying hardware.

The models proposed here are entirely based on combinational circuits and can be analyzed by a standard SAT solver. This is in contrast to previous research in hardware-dependent software verification such as the work of [8] employing SMT solving and the theory of uninterpreted functions with equality, or the work of [18] using explicit unbounded model checking.

Most approaches reported in the literature are based on symbolic execution [3, 17] which is a popular approach to software verification based on symbolically tracking the individual execution paths of a program. However, due to the explicit enumeration of the individual program paths, as pointed out in [2], analyzing the reactive behavior of low-level embedded software with its hardware periphery is very complex. Unlike in many cases of hardware-independent verification, the analysis can no longer be localized to an individual path but the contribution of all possible execution paths must be considered simultaneously. This typically leads to restrictions on the hardware/software interfaces that can be modeled. The work of [8], for instance, restricts the formulation of comparing two assembly programs to programs with very similar control flow graphs (CFGs) that can communicate with the environment only at the beginning and at the end of the execution. In contrast to such a scenario, Sect. 4.4 describes an approach for equivalence checking of programs which are embedded in reactive environments. The approach verifies whether two low-level software programs interact with their environment in exactly the same way, i.e., they produce the same sequences of outputs for every possible input sequence. In order to handle the possibly complex interaction between hardware and software, a *sequence model* of the hardware/software interface is incorporated into our computational models. Then, a *software miter* can be constructed to perform SAT-based equivalence proofs. Experimental results based on the proposed solution are presented in Sect. 4.4.4.

In [12] it is proposed to model the combined hardware/software system in terms of C programs. The approach is based on manually extracting models for the hardware from virtual prototypes in C. Instead of describing hardware abstractly at the software level this chapter presents an approach where the behavior of the low-level software is described by *program netlists* [19] in terms its effects on the concrete hardware implementation. Thus, the approach of [12] can be appropriate to perform early verification of hardware/software systems during design exploration, independently of a concrete implementation. By contrast, the approach described here allows for the verification of concrete implementations of hardware/software designs. This will be explained in Sect. 4.2. The approach has originally been developed to adopt a hardware-dependent but time-abstract view on the software. However, as will be explained in Sect. 4.3, firmware-based design styles may also require a cycle-accurate co-verification of RTL hardware and the firmware. Therefore, in Sect. 4.5 a technique is proposed that bridges two different modeling approaches, namely program netlists, employing path-oriented techniques, as they are traditionally used in software verification [7], and transition-based models used for hardware verification [6, 14]. In particular, the program netlist model is extended to precisely describe cycle-accurate functional behaviors of the processor at its interface, capturing all interactions between hardware and software. This interface model can now be used to connect the transition logic of the hardware at every time point with the software model so that a joint model for hardware/software co-verification is obtained. Experimental results are shown in Sect. 4.5.3.

4.2 Program Netlists

Hardware/software co-verification becomes an important but also difficult task when the software is reactive, i.e., when the tight interaction between the processor executing the software and the surrounding hardware needs to be examined in sufficient detail. A straightforward approach capable of delivering cycle-accurate precision is to model the processor with its program and data memory at the hardware RT level and to use standard hardware verification techniques such as Bounded Model Checking (BMC) [6] to verify properties for this model.

Bounded Model Checking is based on a temporal unrolling of the logic circuitry implementing the finite state machine (FSM) of the design under verification for a finite number of time frames. The resulting computational model is used to check properties expressible as finite sequences of logic relationships. Figure 4.1 illustrates this unrolling. The time frames are gate-level models of the transition function and output function of the design's FSM. A formal property can be expressed as a propositional logic formula over arbitrary signals of the unrolled circuit and can be added as combinational circuitry to the model, (this is not shown in Fig. 4.1). The resulting problem instance is converted to a single formula in conjunctive normal form (CNF) and checked using a SAT solver.

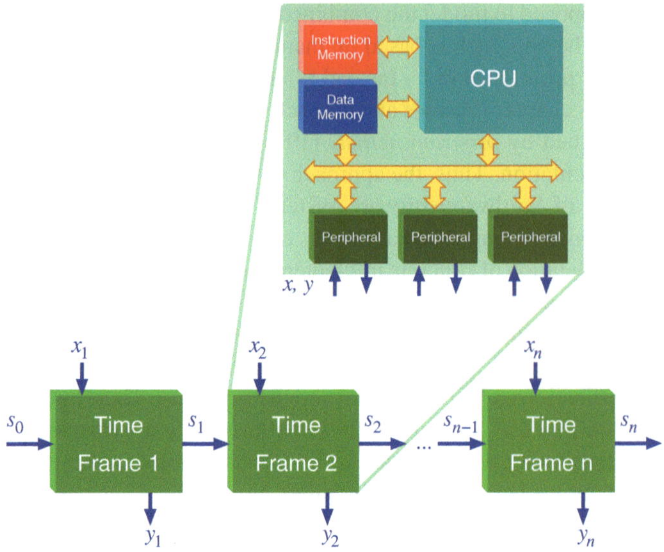

Fig. 4.1 Straightforward BMC-style verification approach

Obviously, this straightforward approach bears complexity challenges. The model includes gate-level representations of the instruction memory with the software, the data memory, the processor and the additional system hardware. Formal properties relating to software can easily span several hundred clock cycles of behavior. However, an unrolled circuit containing several hundred instances of the shown system model is simply not feasible, not even for small processors used in firmware-based design styles (see Sect. 4.5.1). Even if the unrolled circuit can be built and read into the formal proof engine, the computation time for the SAT solver quickly becomes impractical. The reason is that the unrolling implicitly models all possible executions of the software for all possible inputs without any guidance regarding the possible execution traces the software can take. A time frame i represents all possible system states at clock cycle i including the possible states of the program, which, obviously, depend on the possible program states at earlier clock cycles. The SAT solver implicitly enumerates all these possible states for proving the property, and, since it has no guidance of any form, does this very inefficiently. The model can usually not be simplified by constant propagation because a time frame represents not a single location in the program but many possible locations. Approaches relying on a straightforward BMC-style unrolling therefore can work only for small problem instances.

4.2.1 Basic Idea

In [19] a new model for software behavior called *program netlist* has been developed that is compatible with hardware models. It efficiently represents low-level software programs of realistic size that are *reactive* to the hardware, i.e., communication may happen not only at the start and at the end of the program but also continuously during runtime. The model is related to the BMC approach illustrated above, however, some key obstacles to scalability are removed. The basic idea is the following. The unrolling of the processor with its instruction and data memory is not done clock cycle by clock cycle, replicating the full transition function for every time frame, but rather *instruction by instruction*. At branching points in the software the unrolled logic is duplicated, modeling each execution branch separately. This instruction-wise unrolling along execution paths allows for a significant reduction in the amount of logic that needs to be replicated: Since the actual instruction in every unrolled logic block is known and fixed, many constants exist that can be propagated in order to simplify the logic block so that all circuitry that is not needed for modeling the instruction behavior is removed.

In fact, this analysis can be moved to a preprocessing step before unrolling. For a given instruction set architecture and machine program, the behavior of the processor can be precisely modeled for each individual instruction of the program. We call a logic block that models atomically the effects of an individual instruction on a set of state variables an *instruction cell* (IC). The set of state variables that the cell modifies depends on the type of instruction and includes registers from the general-purpose register file, status bits and flags as well as memory locations associated with data variables of the program and input/output registers. (We will discuss the used memory model shortly.) These state variables constitute the *program state* of the programmable hardware/software system. We refer to the subset of these state variables which are internal to the processor as the *architectural state* variables.

Instruction cells are the building blocks of the *program netlist*, i.e., the unrolling of the processor's behavior under the control of the program. We have flexibility in the level of abstraction we choose for modeling the processor. Abstract instruction cells can be used to capture behavior according to the programming model at the instruction set architecture (ISA) level. We can also create instruction cells modeling the concrete behavior of the RTL implementation of a specific processor architecture.

Figure 4.2 shows an example of an instruction cell (modeling a BRANCH instruction). It includes logic circuitry that changes the program state including architecture registers and variables in the data memory. Instruction cells are connected together at the Program State interfaces. Additionally, an instruction cell models the *control flow* of the program using a special state variable called *active*. In the program netlist, all instructions lying on an actual program execution path have their *active* flag set. A BRANCH instruction as shown in Fig. 4.2 produces two possible program states, one for the branch taken and one for the branch not taken. The *active* flag is distributed into the branch selected by the program, as controlled by the instruction logic (signal J in Fig. 4.2).

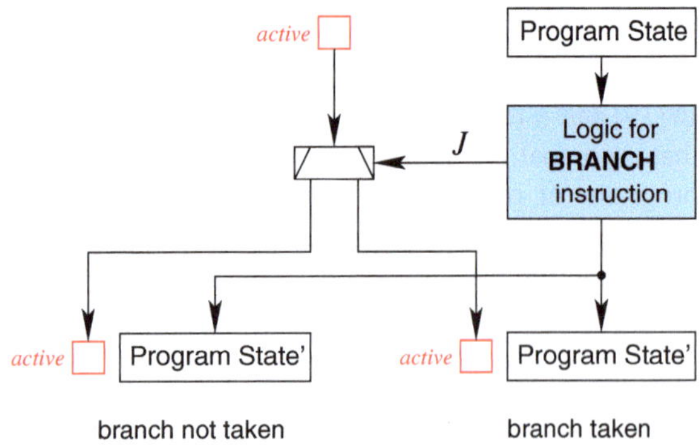

Fig. 4.2 Example: instruction cell for BRANCH instruction

Modeling the control flow in this way is crucial to the performance of SAT-based reasoning on the program netlist as it makes execution paths explicit to the SAT solver. By asserting or de-asserting the *active* signal, whole paths or path segments spanning many time frames can be taken into or out of consideration simultaneously. This gives significant performance improvements over an implicit, unguided enumeration of execution paths as in the straightforward approach discussed above.

Connecting instruction cells together and duplicating paths at branches is, by itself, not sufficient for efficiency because the resulting model can become of exponential size in the number of branches. Instead of building a *tree* of instructions we create a netlist that has the structure of a directed acyclic graph (DAG) with reconvergent paths. As will be shown next, so-called *merge cells* are used for recombining paths in the program netlist in order to avoid exponential growth of the model.

4.2.2 Model Generation

Unrolling of the program involves two steps, as illustrated in Fig. 4.3. We begin with a representation of the control flow graph (CFG) of the program. It is obtained, e.g., from the machine code of the program or from an assembler program. In Fig. 4.3, the nodes in the CFG represent individual instructions. The CFG is unrolled into an *execution graph* (EXG). This execution graph is then used to build the program netlist (PN) by instantiating and interconnecting instruction cells corresponding to the nodes in the EXG.

These two steps are not taken one after the other but instead are carried out in an interleaved fashion. The incomplete program netlist, while it is being built, is used to control the unrolling of the execution graph. The key idea is to determine whether a branch can actually be taken at a particular node in the unrolling. For example, a loop is unrolled until the loop end condition is reached. A SAT solver is used to check whether there exist executions where the active flag of the loop-back branch

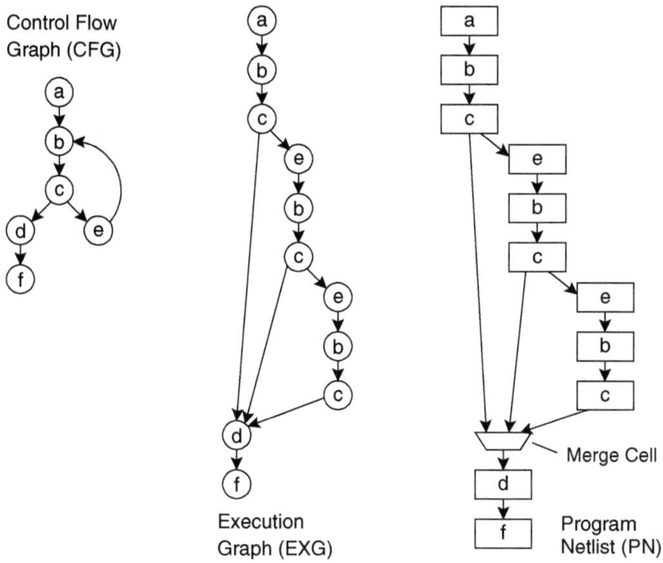

Fig. 4.3 Generating a program netlist

can (still) become active. Similarly, branch destination addresses can be computed on the incomplete program netlist to trace the actual flow of control in the program.

An important component of the model building process is *merging of nodes* in the execution graph. Whenever the control flow modeled in an unrolled path segment reaches a program location that has been visited before, we do not create a new node but instead connect the program state variables to the existing instruction cell for that location, provided this does not introduce a cycle in the graph. This is done using multiplexers in the program netlist called *merge cells* (see example in Fig. 4.3). Merging keeps the model compact by sharing of sub-graphs and produces a DAG with reconvergent paths (rather than a tree). Note that the sub-paths being merged in a merge cell can never be active simultaneously. This is guaranteed by construction of the program netlist with *active* flags.

4.2.3 Modeling Memory and I/O

Efficiently modeling accesses to the data memory and to the environment (e.g., to HW peripherals) is crucial for the scalability of the method. The left-hand side of Fig. 4.4 shows an instruction cell modeling a read or write access to data memory. The shown selection logic creates an access path from the *Load* or *Store* instruction cell for the given address, *addr*. In case of a *Load*, the architectural state is modified with the selected *data*. In case of *Store*, the data memory is updated with *data* from the

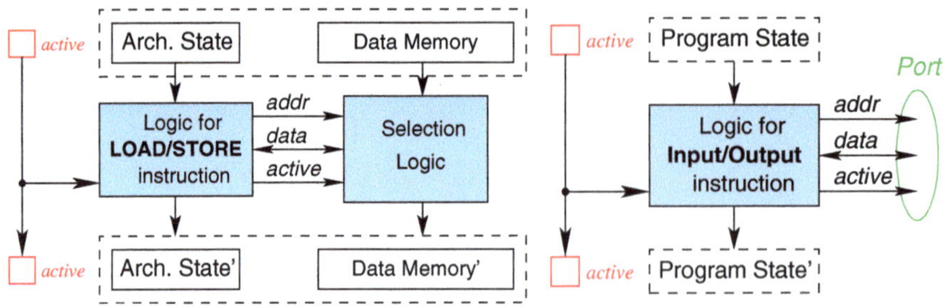

Fig. 4.4 Instruction cells for memory accesses and environment I/O

architectural state. Modification is enabled if the *active* flag is set. In order to keep the selection logic block compact, a combined simulation/SAT-based algorithm [5, 20] is used to compute the set of addresses the *Load/Store* instruction at the given program location can actually access. This reduces the size of the selection logic greatly because only few data memory locations need to be multiplexed for a given instruction cell. Although, in principle, the range of addresses accessed by a program can be huge, in the intended application domain for our technique (low-level, hardware-dependent software) the number of addresses used by the software is usually limited and restricted by design.

This representation of the program state can be viewed as an associative memory model like the one proposed in [9] for formal verification of assembler code by bounded model checking. Note, however, that in our approach the associated memory entries are created statically during model generation and not dynamically at proof time through value assignments in the reasoning engines. The resulting logic in the program netlist can be much simpler and can lead to more efficient SAT reasoning during verification. This is a direct benefit of the instruction-wise and path-oriented unrolling of the software in the program netlist as opposed to a BMC-style unrolling of the processing hardware as in [9].

The right-hand side of Fig. 4.4 illustrates how input/output is modeled using instruction cells. The input/output instruction cell provides an interface for the program netlist called *port*.

Definition 4.1 The *port* of an input/output instruction cell *i* is a set of three logic signals called *i.addr*, *i.data* and *i.active*. □

The *addr* signal provides the address of an environment location, e.g., a peripheral device register. The *active* flag indicates when an access occurs and data is transferred through the *data* signal. Using such ports, the sequences of input and output accesses of the software along different execution paths can be modeled. In Sect. 4.4 ports and access sequence models are used to create a *software miter* for equivalence checking of reactive low-level software. In the same way as for memory accesses, a simulation/SAT-based algorithm is used to compute the possible addresses an input/output instruction can access [20]. This is used in Sect. 4.4 for equivalence checking to reduce the amount of logic necessary to model all possible access sequences exhibited at the hardware/software interface. It is also used in Sect. 4.5 for keeping the logic needed for a cycle-accurate input/output model as small as possible.

The program netlist model obtained in this way allows for efficient SAT-based reasoning in applications like equivalence checking (Sect. 4.4) or property checking [4] for low-level embedded system software. Key to efficiency is the fact that most of the control flow related information is computed beforehand and built into the model. The program netlist is an explicit representation of all possible execution paths in the software, while the data path information is still contained implicitly in the combinational circuitry inside the instruction cells. This makes the model particularly amenable to SAT-based proof algorithms.

4.3 Verification Scenarios for HW-dependent Software

The kinds of computational models used for hardware-dependent software verification rely on how the examined hardware and software components are actually integrated into the system. In the following, two main hardware/software integration scenarios are described and in Sects. 4.4 and 4.5 it is presented how formal verification can be performed for each scenario.

The first scenario is used traditionally in SoC design flows. Processor cores are integrated into the hardware system as components of a CPU bus. They usually communicate with the rest of the system in a speed-independent way using some bus protocol with handshake mechanisms in order to accommodate for access latencies. Speed-independent communication is key since the execution time in pipelined processors, especially when advanced architectures based on out-of-order execution are employed, is difficult to predict. Similarly, caches have a difficult-to-predict timing behavior and provide another reason for speed-independent bus communication.

Assuming the correct implementation of the bus protocol, as can be verified by standard techniques of formal hardware verification, it is possible to model the software in a time-abstract way. This is exploited in the program netlists of Sect. 4.2 by creating time-abstract instruction cells. In this way, even for complex processor architectures compact models can be obtained.

It is important to note that even though the concrete timing, in terms of HW clock cycles, is abstracted away from the program netlist, the original ordering of the instructions during execution is preserved in the model. This characteristic of the program netlist is particularly important when analyzing the hardware/software interface of reactive programs. Reactive software communicates with the environment continuously at distinct time points and the ordering in which the exchange of information takes place is crucial for functional correctness of the system behavior. For example, for the case of equivalence checking, as will be shown in Sect. 4.4, it needs to be proven that two different programs interact in the same way with the environment. Therefore, verification needs to consider not only the data values exchanged with the environment but also the ordering of the data exchange.

Apart from equivalence checking, also property checking for time-abstract scenarios using program netlists has been researched in [19].

Besides conventional design styles where software and hardware are integrated by employing CPU buses as explained above, there are also new design approaches for which a clock cycle-accurate analysis is required. This introduces a second scenario which is described in the following and motivates the extensions to the program netlist model, presented in Sect. 4.5.

When designing SoCs it has become increasingly popular to replace dedicated hardware components by a number of simple processor cores whose timing behavior is fully predictable. For example, processors in the style of the Intel 8051 or the Xilinx PicoBlaze are popular in ASIC-based and FPGA-based design flows, respectively. Such a design style offers advantages:

1. The design time is reduced and product updates can be made more easily by making changes in the software.
2. Especially for FPGAs, due to a well-optimized design of the processor, the resulting implementation may need less chip area when compared to a conventional implementation with standard synthesis.

The software running on the instantiated cores is usually not meant to be visible to the users. It is provided as firmware with the design and may be loaded into a ROM.

In many cases, the cores are not directly connected to the rest of the system using (standardized) communication interfaces like buses but instead are embedded into the surrounding system using special interface hardware, sometimes called "wrapper RTL". Such design styles allow for a tight integration delivering high performance because the exact timing of the processor hardware and its software are known at design time. An SoC module designed in this way is shown in Fig. 4.5. The module consists of two processor cores tightly integrated with their firmware, wrapper RTL and some additional hardware.

For such an ad hoc design style, verification is important because the hardware/software interface is usually custom-designed and, thus, error-prone. As pointed out in [21], traditional verification approaches based on instruction set hardware/software co-simulation would never fully capture the entire hardware and firmware system behavior in one tool environment. Verifying the firmware in isolation, while possible, would require a hardware bus-functional model interface. This makes the simulation of such firmware-based IPs complicated and creates the need for an additional behavioral test bench [21]. Similarly, also a formal approach that verifies hardware and software in separation would require the tedious task of modeling the interface between them by a set of constraints.

Therefore, we propose a formal co-verification approach instead. The program netlists of Sect. 4.2 are very attractive for this purpose because they can be generated completely automatically. On the other hand, due to their abstract, non-cycle-accurate nature, they cannot be directly integrated into the RTL descriptions of the hardware and they do not allow for a cycle-accurate analysis. In Sect. 4.5, we therefore present extensions to make program netlists cycle-accurate and show how to create a joint model for formal co-verification of hardware and firmware.

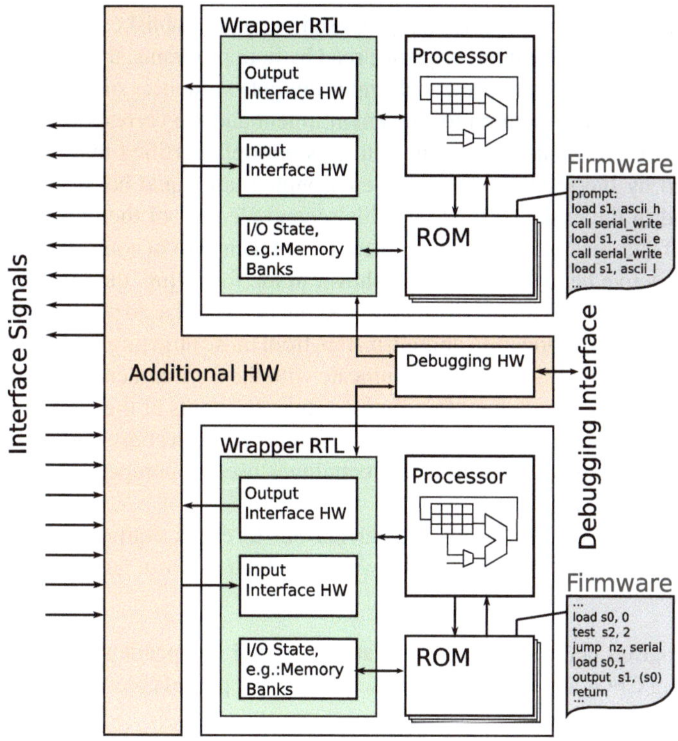

Fig. 4.5 Firmware-based IP core

4.4 Equivalence Checking of HW-dependent Software

During design of an embedded system, the software usually undergoes several transformations. For instance, embedded software is frequently optimized, automatically or manually, in order to meet design requirements on program execution speed, memory footprint (code size), and power consumption. Furthermore, often during the design or field time of a product, features are added to a given software, extending the existing functionality. Finally, embedded software is also often ported to different hardware platforms. For all these cases, equivalence checking is a valuable tool since it can be used to certify that the original functionality of the software is not damaged or altered by the applied transformations.

In this section, a fully automated method to formally prove the functional equivalence of hardware-dependent programs is presented. We focus on describing the main ideas, originally presented in [20], that enable building an efficient computational model called *software miter* to solve the equivalence checking problem for reactive programs. Additionally, we extend the concept of data sequences to model also *address sequences*. This extension allows us to verify, when required, the address interleavings of the software when accessing different data environment locations.

The equivalence criterion for the proposed method establishes that two programs are equivalent if for any input sequence read by both programs, the output sequences produced by the programs are equal. Input (output) sequences of a program contain the values read from (written to) the environment and the corresponding orderings. According to this equivalence criterion, it is not only certified that the data values exchanged by the programs with the environment are equal but also that data are exchanged in the exact same order. This second element of the proof is especially relevant if the reactiveness of the software is taken into account. For taking into account reactive behavior, as will be shown in the following, the program netlist is extended with a global model of the input/output behavior based on access sequences.

Notice that our application domain differs from those targeting conventional transformational algorithms which communicate with the environment only at two points in time, i.e., at the beginning of execution when the inputs of the algorithm are read and at the end of execution when outputs are returned. As a consequence, the method presented here differs from previous techniques used to compare transformational software such as [16].

The following general steps are carried out to check equivalence of two programs *G* (for "golden") and *R* (for "revised") which may run on different hardware platforms:

- The program netlists for *G* and *R* are generated independently from the corresponding CFGs and instruction cells. For this, the process presented in Sect. 4.2 is followed.
- Each program netlist is extended with a sequence-based input/output model. Section 4.4.1 presents more details on how this model is constructed.
- A software miter is built by using the program netlists, the sequence-based input/output model and a bijective mapping of input/output addresses from *G* to *R* provided by the user. To construct the software miter we take advantage of the fact that program netlists are compositional. Section 4.4.2 gives more details on this step.
- Finally, a decision procedure (e.g., a SAT solver) is called iteratively to prove the equivalence of each mapped sequence. In Sect. 4.4.3, we show how incremental SAT solving techniques can be employed to reduce verification run time.

4.4.1 Sequence-Based Model of the HW/SW Interface

This section presents a model of the hardware/software interface which employs the concept of input/output sequences to represent the exchange of data of the software with its environment. As mentioned before, since the software is embedded in a reactive environment, it is necessary to include in the model a representation of the exchanged data values as well as the orderings describing the timing of the data exchange. A good example of a reactive program is a software-implemented bus agent. According to the bus specification, the bus agent transfers frames of information ensuring that the transmission order of the individual fields composing a frame is preserved.

From the software perspective, communication with the environment takes place at input/output CPU instructions accessing environment locations which belong to the system's address space. In memory-mapped input/output systems, typically load/store instructions transferring information from/to device registers in hardware peripherals or IP cores serve this purpose. Another example can be a hardware device controlled by a dedicated CPU port. The interface model presented here is independent of the kind of mechanisms used by the CPU to communicate with the environment.

We denote the set of data environment locations accessed by a given program by $A = \{a_1, a_2, \ldots, a_m\}$. Each a_j is an address of such a location. These locations are read (input data location) or written (output data location) by the software to communicate with the rest of the system. They may correspond to registers in hardware devices or to shared memory locations used for exchanging data with other software components or layers (e.g., the application layer). Note that A is actually a subset of the set of all data locations accessed by the software. It contains only data locations which are relevant to the external input/output behavior. Data locations corresponding to software variables in main memory that are not externally visible are not contained in A. According to this, the model of the hardware/software interface describes the sequences of accesses performed by the software only to each of the data locations in A.

In order to represent the ordering of accesses in the sequence model for each data location $a_j \in A$ a new temporal variable t_{a_j} is added to the program netlist. This variable represents the position (index) of a particular element in the sequence of accesses to a given data location a_j. We can think of it as an abstract access time point. In the program netlist these time variables are propagated and updated uniquely at input/output instruction cells which access a particular data location a_j.

From a software perspective, the hardware/software interface is directly affected by input/output instructions of the software, i.e., instructions that access the data locations in A. Therefore, the first issue that needs to be solved in order to model the hardware/software interface consists of identifying the set of input/output instructions of a given program and of determining which of the elements of A are actually accessed by each input/output instruction. This issue can be solved easily because after program netlist generation, the complete address space accessed by the software is known. As stated in Sect. 4.2.3, for each input/output instruction the values taken by the *port* signal *addr* (cf. Definition 4.1) are computed during generation of the program netlist by using simulation together with enumerative SAT [20]. Therefore, when building the sequence model, it is only necessary to check for each instruction cell of the program netlist if the values of *addr* signal (cf. Fig. 4.4) belong to the set of data locations, A.

With the previous information, we define then the set of all input/output instruction cells that can access the data location $a_j \in A$ as $I_{a_j} = \{i_1, i_2, \ldots, i_{n_j}\}$. Once input/output instruction cells are identified then the logic for them is extended with incrementers for the abstract time variable as shown in Fig. 4.6. For instance, if a

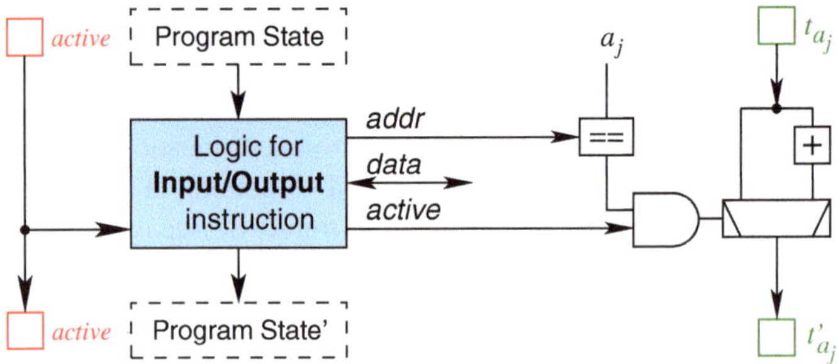

Fig. 4.6 Logic for updating time variables used for data sequences

given instruction cell can access the location a_j, then, if the instruction cell is active (i.e., the access is performed), the value of the temporal variable t_{a_j} is incremented by one. We denote the value of the temporal variable at the instruction cell i by $i.t_{a_j}$.

Based on the temporal variables we can construct the logic for each sequence element. The k-th written data value, denoted by $data_{a_j}(k)$, to a given location a_j, is described by the following if-then-else construct, built for the set $I_{a_j} = \{i_1, i_2, \ldots, i_{n_j}\}$ of all input/output instruction cells that are able to access a_j:

$$\begin{aligned} &data_{a_j}(k) := \\ &\qquad\quad \textbf{\textit{if}}\ (i_1.active\ \textbf{\textit{and}}\ i_1.addr = a_j\ \textbf{\textit{and}}\ i_1.t_{a_j} = k)\ \textbf{\textit{then}}\ i_1.data \\ &\quad \textbf{\textit{else if}}\ (i_2.active\ \textbf{\textit{and}}\ i_2.addr = a_j\ \textbf{\textit{and}}\ i_2.t_{a_j} = k)\ \textbf{\textit{then}}\ i_2.data \\ &\qquad\quad \ldots \\ &\quad \textbf{\textit{else}}\ i_{n_j}.data \end{aligned}$$

The logic for $data_{a_j}(k)$ builds a cascade of multiplexers where each multiplexer is connected to a single input/output instruction cell (belonging to I_{a_j}). The selection signal of each multiplexer is set to *true* if three conditions are met, namely, (1) the instruction cell is active, (2) location a_j is addressed and (3) the time variable has the value k. The last two conditions are necessary because an instruction cell can access more than one data address and, also, can perform the accesses at different abstract time points. When the select signal of a given instruction cell's multiplexer is *true* then the data value of the sequence at time point k is assigned the value of the data signal.

The logic for $data_{a_j}(k)$ can be simplified because normally for a given sequence element k just a single instruction cell or a small subset of I_{a_j} can actually access the interface at the abstract time point k [20]. Therefore, not all input/output instruction cells belonging to the set I_{a_j} need to be considered in the logic for $data_{a_j}(k)$ and consequently the amount of multiplexers can be reduced. As an example, assume that a program with the CFG of Fig. 4.3 writes to an environment location at instruction b. From the program netlist obtained (shown on the right hand side of Fig. 4.3), it can be seen that the longest access sequence for the location accessed by instruction b has three elements. This happens when the program takes the right-most path through the execution graph, visiting the b-instruction three times. For each element of the access sequence there is only a single b-instruction cell that can write to the location.

We implement an EXG traversal algorithm for identifying the instruction cells that can access a particular location a_j at a given time point k. The algorithm takes as input the EXG which is topologically sorted and the information about the reachable address space of the software. For a given location a_j, the algorithm propagates sets of access count values, i.e., sets of possible values of the index variable t_{a_j} as introduced above, through the EXG in topological order beginning at the root node(s). Whenever an instruction cell is visited that accesses address a_j, every access count in the set is incremented. The set represents the possible indexes of the elements in the output sequence for a_j that are affected by the instruction. For instructions that do not access address a_j, the set of access counts is propagated without modification. After the traversal, we can compute, for each sequence element k, the set of instruction cells, $\tilde{I}^k_{a_j} \subseteq I_{a_j}$, that affect k. This optimization not only reduces the size of the sequence model but also speeds up verification run time since the decision procedure does not waste time anymore analyzing and eventually discarding input/output instruction cells which are now known to be irrelevant for a particular sequence element.

Another auxiliary signal also included in the sequence model is $active_{a_j}(k)$. It is asserted in any execution path where the k-th read or write access to a data location a_j occurs. As will be explained later, this signal helps to speed up the verification by ensuring that only the execution paths related to the k-th access sequence point are considered by the decision procedure in a particular proof. All other paths (if any) related to other time points will be disregarded by the solver.

$$
\begin{aligned}
&active_{a_j}(k) := \\
&\qquad \textbf{\textit{if}}\ (i_1.active\ \textbf{\textit{and}}\ i_1.addr = a_j\ \textbf{\textit{and}}\ i_1.t_{a_j} = k)\ \textbf{\textit{then true}} \\
&\quad \textbf{\textit{else if}}\ (i_2.active\ \textbf{\textit{and}}\ i_2.addr = a_j\ \textbf{\textit{and}}\ i_2.t_{a_j} = k)\ \textbf{\textit{then true}} \\
&\qquad \ldots \\
&\quad \textbf{\textit{else if}}\ (i_{n_j}.active\ \textbf{\textit{and}}\ i_{n_j}.addr = a_j\ \textbf{\textit{and}}\ i_{n_j}.t_{a_j} = k)\ \textbf{\textit{then true}} \\
&\quad \textbf{\textit{else false}}
\end{aligned}
$$

While the previous analysis describes the sequences of data exchanges for single data locations, in some cases it is also important to consider the interleaving of accesses to different data locations. For example, for the bus agent introduced above, it can be required that the initialization of the surrounding hardware takes place before the data payload can be transferred, otherwise the hardware periphery is not ready to communicate with the agent. Therefore, verification needs to prove that initialization is executed before the data payload transmission takes place. For this purpose, a model for address (data location) sequence can be employed. For the bus agent example, the first element of the address sequence should equal the value of the address used for initialization and the rest of the elements of the sequence should correspond to the address used for transferring the payload and so on. The model generation for the address sequence follows the same strategy based on incrementers as was introduced for the data sequences (see Fig. 4.6). The logic added for updating time points of address sequences is shown in Fig. 4.7. This logic is again added to

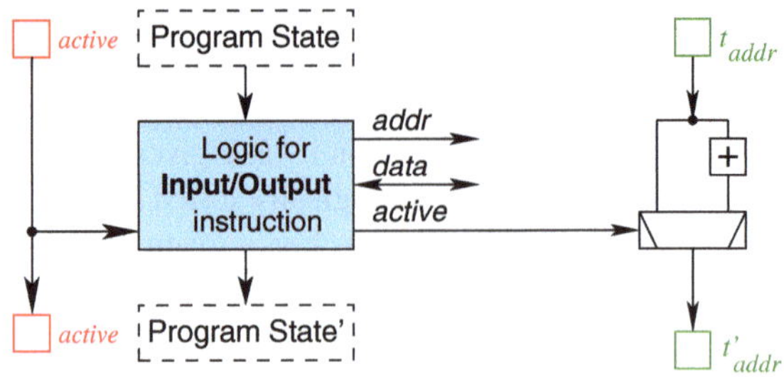

Fig. 4.7 Logic for updating time variables used for address sequences

all input/output instruction cells of the program netlist. However, note that for this case, comparators are not required as all environment locations in the set A need to be considered simultaneously.

Likewise, the corresponding logic for each element of the address sequence is described as follows.

$$\begin{aligned}
&addr(k) := \\
&\qquad\quad \textbf{\textit{if}}\ (i_1.active\ \textbf{\textit{and}}\ i_1.t = k)\ \ \textbf{\textit{then}}\ i_1.addr \\
&\quad \textbf{\textit{else if}}\ (i_2.active\ \textbf{\textit{and}}\ i_2.t = k)\ \ \textbf{\textit{then}}\ i_2.addr \\
&\qquad\quad \ldots \\
&\quad \textbf{\textit{else}}\ i_{n_j}.addr
\end{aligned}$$

The functions for the signals $data_{a_j}(k)$, $active_{a_j}(k)$ and $addr(k)$ encapsulate the interface of the software with the environment and are next used for solving the equivalence checking problem.

4.4.2 *Software Miter*

The previous model of the hardware/software interface makes it possible to formulate the *equivalence* of hardware-dependent programs in a straightforward way as follows.

Consider two low-level hardware-dependent programs G and R. For these programs, the inputs and outputs are defined by the user as sets of input data locations X_G, X_R and output data locations Y_G, Y_R.

The user provides additionally a bijective mapping that assigns to every input data location $x_G \in X_G$ of program G an input data location $x_R \in X_R$ of R. Also, a bijective mapping assigning elements of Y_G to elements of Y_R must be provided. Then, the program netlists for G and R and the corresponding sequence models are created with respect to the user-defined environment locations. Finally, the two program netlists together with their corresponding interface models are combined as follows.

Mapped inputs are set equal by connecting every input sequence element $data_{x_G}(k)$ of program G with the corresponding sequence element $data_{x_R}(k)$ of program R.

This ensures that the input assignments of the programs are equal as established in the equivalence criteria defined at the beginning of this section. At this point, it is expected that the sequence lengths are the same for both programs. If this is not the case, then no sequence mapping is possible and the programs are not equivalent.

Similarly, for each sequence element of the mapped outputs $data_{y_G}(k)$, $data_{y_R}(k)$ and their respective active signals $active_{y_G}(k)$, $active_{y_R}(k)$ the following set comparisons are implemented.

$$\begin{aligned}
&\text{equiv}(y_G, y_R, k) = \\
&(active_{y_G}(k) = active_{y_R}(k)) \; \boldsymbol{and} \\
&(active_{y_G}(k) \; \textbf{\textit{implies}} \; data_{y_G}(k) = data_{y_R}(k))
\end{aligned} \tag{4.1}$$

The function $\text{equiv}(y_G, y_R, k)$ can be seen as a property, in particular a Boolean predicate. The first condition in Eq. 4.1 expresses that a sequence element must be produced by both programs under exactly the same input conditions. Remember that both, $active_{y_G}(k)$ and $active_{y_R}(k)$, are asserted for exactly the input conditions that make the respective program, G or R, produce the output sequence element k. They are deasserted if k is not produced. The second condition states that whenever the output sequence element k is produced then its data value must be the same in both programs.

For checking equivalence of the address interleavings the same kind of comparison can be implemented for the sequence elements $addr_G(k)$ and $addr_R(k)$ of the programs G and R, respectively.

The final model results in a software miter with a set of mapped inputs and a vector of comparison for the outputs. It must be checked whether or not all of these comparison outputs always yield *true*. If this is the case then both programs G and R are equivalent.

4.4.3 Equivalence Checking Using SAT

In order to compute the proofs expressed by Eq. 4.1, we take each pair of mapped outputs (y_G, y_R) and call the SAT solver iteratively for all related sequence points as shown in Algorithm 4.1. As for the inputs, sequence lengths for the outputs are also expected to be the same for each environment location, i.e., for the mapped locations y_G, y_R, the corresponding sequence lengths m_G, m_R must have the same value.

for $k \leftarrow 1$ **to** m_G **do**
| $SAT - prove$: $\text{equiv}(y_G, y_R, k)$
end

Algorithm 4.1: Iterative SAT proofs for a given location

For each iteration k the SAT solver enumerates all involved execution paths by using the *active* signal mechanisms mentioned in Sect. 4.2.1. The Eq. 4.1 places Boolean constraints on the *active* signals of the compared output sequence elements. By taking decisions and propagating values into the *active* signals of the elements of the program netlist, the SAT solver explores the consequences of these constraints on the control flow of each of the programs, G and R. The SAT search is, thereby, guided to consider only those execution paths that are related to the equivalence check of the currently considered output sequence element k. By construction of our models, the *active* flags in both program netlists are assigned by the SAT solver such that only related execution paths in the two program versions are considered simultaneously. Multiple executions paths can be implicitly considered at once. Clauses can be learned that express relationships between corresponding execution paths in both programs. All of this helps to significantly enhance the efficiency of the SAT proof.

It is clear that the cone of influence of the proofs in Algorithm 4.1 grows incrementally with k because, if an access happens at sequence index $k + 1$ then an access to the same port has happened at sequence index k in the same execution path. This means that the cone of influence of the constraint $\text{equiv}(y_G, y_R, k + 1)$ contains the cone of influence of the constraint $\text{equiv}(y_G, y_R, k)$.

We can take advantage of this fact and employ *incremental SAT* techniques to reuse the knowledge acquired by the SAT solver when proving $\text{equiv}(y_G, k)$ for the subsequent proof of $\text{equiv}(y_G, k + 1)$. The individual points in the sequence of equivalence proofs can be seen as "internal equivalences" for all later proofs and have a similar speed-up effect as internal equivalences in combinational hardware equivalence checking.

4.4.4 Experimental Results

The following experimental evaluation demonstrates that it is feasible to perform equivalence checking of industrial hardware-dependent programs using the approach presented above.

The concepts described in this section have been implemented within the formal verification environment called FCK (*Formal Checker Kaiserslautern*). For building the software miter as explained in Sect. 4.4.2, FCK starts with generating the program netlists for every program to be handled (golden and revised versions) from the corresponding machine codes. While generating the program netlist (preprocessing phase) also the memory model (Sect. 4.2.3) is built. Subsequently, the hardware/software interface model is constructed (Sect. 4.4.1) for each of the program netlists. Based on the mapping information provided by the verification engineer comparison functions and verification targets are generated. If required, input mapping and comparison functions can be adjusted by the user. Finally, the equivalence proofs are computed using an incremental SAT solver (MiniSAT [10] in the current implementation). In case of a bug, FCK presents a counterexample as a pair of active program traces showing a scenario in which both programs behave differently. For every trace, the tool presents the values of the program states along active paths on the program netlist corresponding to a given input assignment of the mapped inputs. Note that there can be no false counterexamples because the program netlists in the software miter exactly represent all execution paths beginning at the initial states of the programs and there are no approximations of state sets in the model.

The following experiments are based on two relevant examples of hardware-dependent software, namely, an industrial software implementation of the automotive protocol LIN and a serial synchronous interface. Both examples are mainly control driven.

All experiments are performed on an Intel Xeon E5420 CPU at 5 GHz with 16 GB RAM.

4.4.4.1 LIN Driver

The considered driver software was originally developed by Infineon Technologies AG and implements the LIN bus protocol for a master node. For these experiments it was adapted to run on the open-source 32-bit five-stage pipelined processor Aquarius [1]. The driver comprises about 1350 lines of hardware-dependent, low-level C code and inline assembly. It can be configured such that transmission and reception modes are allowed, data-length is variable up to 8 bytes, and the used IDs can be modified. The driver interacts with the LIN bus by means of a UART containing status, configuration and data registers. The UART is accessible as a memory-mapped input/output device. The driver also interacts with a user application via shared memory consisting of the received data, the data to be transmitted and additional status information (e.g., the status of the transmission). All program netlists generated for the driver model these software features.

The described equivalence checking approach was applied to this software, considering different scenarios in which code is subjected to automated and/or manual transformations. The GCC compiler was used applying three different optimization levels to the source code, starting from level zero (LIN 10, cf. Table 4.3) with no optimizations being activated and increasing the aggressiveness of the compiler optimizations up to the maximum level two (LIN 12). Engineering changes were introduced into the code in different parts of the program (LIN modif.). For all these cases the program versions were verified to keep the same input/output behavior. Experiments were also conducted with a version of the driver containing an error in the computation of the checksum (LIN error). This code was obtained based on the code version modified by engineering changes (LIN modif.) and by making further manual changes that introduced an error.

Table 4.1 shows the times required to generate each program netlist (preprocessing phase). These include the times necessary to explore the address spaces accessed by each instruction cell that interacts with the environment or with data memory.

Table 4.1 CPU times for model generation

Program	CPU time (s.)	
	SAT only	SAT and constant propagation
LIN (10)	6828.4	26.8
LIN (11)	1087.8	12.4
LIN (12)	1016.9	11.6
LIN (modif.)	1323.0	12.9
LIN (error)	1298.2	12.1

As can be observed, constant propagation drastically reduces the CPU times. This confirms that the memory addressing mechanisms, as they are employed here and in similar applications, can result in a large number of constant address values so that the generation of program netlists remains tractable even at the presence of a large address space.

Before calling the SAT solver it was checked that all versions of the LIN driver have the same number of access sequence points. This was a first indication that the programs are equivalent in all cases. Table 4.2 presents information on the interface model. Times to build the access sequence, as explained in Sect. 4.4.1, were negligible since only a simple graph traversal of the EXGs are needed in order to identify input/output instruction cells.

For each output sequence point a SAT check was then performed. Data for the software miters and the proof times are shown in Table 4.3. In all cases, except for the comparison LIN (11) versus. LIN (error), the programs were proven equivalent

Table 4.2 Program netlists: interface model

Program	No. locations		No. seq. points	
	input	output	input	output
LIN	6	5	25	42
SER	2	4	992	4

Table 4.3 Equivalence checking: proof results

Golden	Revised	Miter size (inst. cells)	Proof time (s.)	Memory usage (MB)
LIN (10)	LIN (11)	13764	692.3	777.5
LIN (10)	LIN (12)	13198	766.2	698.1
LIN (11)	LIN (12)	11520	419.5	343.0
LIN (11)	LIN (modif.)	11904	500.5	470.5
LIN (11)	LIN (error)	11915	295.1	336.2
SER (orig.)	SER (ported)	11760	188.2	404.6

according to our formulation. For the equivalence proof of LIN (11) versus. LIN (error) a counterexample was returned by the SAT solver. This counterexample was composed of two *active* execution traces: one for LIN (11) and the other for LIN (error). The counterexample presents an input assignment to the mapped inputs of both programs which, in the considered case, produced a mismatch of the values written to the UART's transmission buffer. The value written to the UART buffer corresponded to the checksum field of the LIN-protocol and, specifically, it could be observed that the erroneous behavior occurred at the 12th time point of the output access sequence belonging to this buffer.

For the proofs the technique described in Sect. 4.4.3 was employed. By using internal equivalences detected by incremental SAT, CPU times could be reduced to about 36% on average.

4.4.4.2 Serial Synchronous Interface

The interface implements a serial synchronous receiver using a round-robin scheme that iteratively samples a clock synchronization signal and a data-input serial line. In every transfer the data is passed byte-wise to the user application until a 32-bit word has been received. In order to guarantee a finite unrolling a model generation constraint was added that limits the number of sampling actions to ten (five for each clock-phase).

The code was initially developed for the Aquarius (SER (orig.)) and was later ported to run on the ARM7-TDMI architecture (SER (ported)). Then, the equivalence of both versions of the code was formally proven.

The serial synchronous interface provides an interesting case study since it contains a complex nested-loop structure with a high number of branches. On the other hand, when compared to the LIN driver, this program has low traffic with data memory. Therefore, the times for the model generation in this case were dominated by the checks on the *active* signals performed for path pruning at every branch during the unrolling. The times for model generation were: 698.4 s for SER (orig.) and 645.0 s for SER (ported).

Table 4.2 shows information on the interface model. Both versions of the serial interface presented the same number of access points on the interfaces. Input sequence points correspond to the individual samples of the serial data line and of the clock signal. Output points of the interface relate the corresponding storing accesses of the received data to the user application.

Table 4.3 presents the information on software miter construction and proof times. The programs were proven to be equivalent. Due to incremental SAT run times were reduced to 27%.

4.5 Cycle-Accurate HW/SW Co-verification of Firmware-Based Designs

4.5.1 Joint Hardware/Firmware Model

This section describes a model for hardware/software co-verification of firmware-based designs. As shown in Figs. 4.1 and 4.5, systems are composed of processor cores, their surrounding hardware, also called "uncore" hardware, and the software to be executed. In the following analysis, the uncore hardware (wrapper RTL, peripherals and relevant IP components) should be distinguished from the core hardware (CPU, data memory, instruction memory, and communication infrastructure) with its software. For the sake of a simple terminology, in the following, we use the term *hardware* only for the uncore parts of the system. In our hardware-dependent software view, since the software behavior is described completely in terms of the core hardware, the term *software* or *firmware* subsumes not only the considered program but also the core hardware on which it is running.

Every core architecture contains input/output instructions that interact with the uncore hardware. In our model, they are represented by input/output instruction cells equipped with a *port* as introduced in Definition 4.1. The model to be presented precisely describes the functional behaviors of the system cycle by cycle over finite-time windows. For taking both the hardware and the software into account, an approach is taken that combines two different kinds of unrollings as depicted in Fig. 4.8.

On the one hand, the uncore hardware is unrolled in a classical BMC [6] fashion by instantiating a copy of the associated transition relation at every time step (lower part of Fig. 4.8). On the other hand, the unrolled software is modeled by a program netlist,

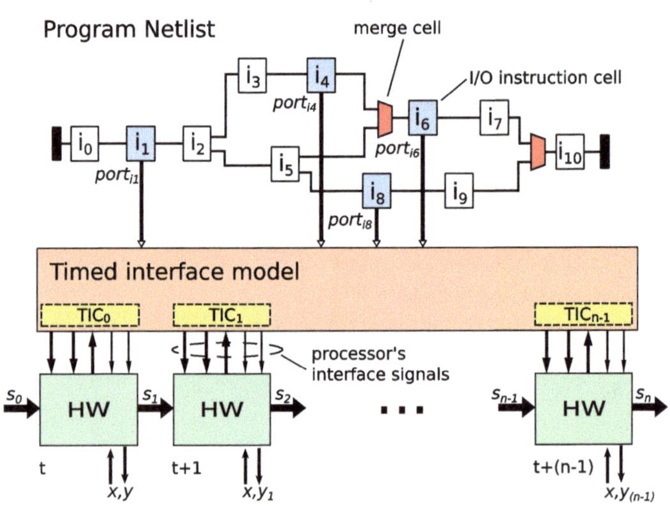

Fig. 4.8 Mixed unrolling approach for efficient HW/SW modeling

instruction by instruction, representing all possible executions of the programmable system (upper part of Fig. 4.8). Time granularity in this part of the model is given by processor instructions and not by clock cycles.

For constructing the hardware/software model, we take advantage of the fact that a program netlist can be instantiated as a hardware component and can be extended with a new model of the processor's interface that allows us to combine the program netlist together with the hardware into a single model, as shown in Fig. 4.8. For this purpose, different simplifications are performed in order to reduce the amount of logic required to model the processor's interface and to ease the reasoning on the resulting composed model. In particular, an algorithm (to be presented in the next subsection) has been developed to statically determine the subset of input/output instructions that can interact with the relevant system's hardware at a specific time point (clock cycle) of the unrolling. With this information the model of the processor interface can be reduced because the new modeling logic depends only on these relevant input/output instructions and not on other instructions not accessing the interface of the processor or addressing memory spaces that do not correspond to the relevant uncore hardware. Since this information is explicitly added to the model, the decision procedure used to reason on the model can directly exploit this information instead of deducing it from another more complex representation.

In general, developing such an algorithm may appear complicated. However, for the approach proposed in the next subsection, it turns out doable since the required algorithm can employ information readily available as a result of the generation procedures for program netlists. In particular, when determining the set of instructions that can be executed at a specific time point of the unrolling, the algorithm benefits from working directly on the execution graph which contains explicit information about the control flow of the program and its corresponding accessed memory address space.

4.5.2 *Timed Interface Model*

Most of the modeling challenges of creating a combined hardware/software model stem from the combination of two unrolling styles with different temporal resolution (Fig. 4.8). Since the hardware is unrolled in a cycle-accurate manner, state variables and, in particular, signals connecting to the processor are already contained in the model at every modeled time point. For the software, however, the situation is different because the program netlist as presented originally in [19] is a time-abstract model. As explained in Sect. 4.2, instruction cells atomically represent how a given ISA instruction modifies the program state, abstracting from any intermediate steps carried out by the CPU during instruction execution. This is also true for input/output instructions (Fig. 4.4), which do not model how the interaction between the processor and the hardware specifically takes place in time. In the same way, even though a program netlist represents sequences of instructions executed by the processor, the state of a program at a particular absolute time point is not known, because it depends on the inputs of the program and the execution path actually taken.

These issues can be resolved by adding a cycle-accurate model of the processor's interface to the overall model that accurately represents the interface signals of the processor for each time point of the unrolling. In the following, we show in detail how such an interface model can be constructed by (1) adding new abstractions describing the behavior of the processor's interface and by (2) creating additional resolution logic for deciding the time points and the values communicated between hardware and software.

4.5.2.1 Timed Interface Cells

We define a *timed interface cell* (TIC) as an abstract model representing the state of the signals belonging to the processor's interface at a particular time point. TICs (the yellow boxes in the "timed interface model" in Fig. 4.8) are hardware-dependent models specific to each processor architecture. TICs can be classified into DATA-TICs and IDLE-TICs. DATA-TICs transport input/output information such as data, addresses and control values between the software and the relevant system hardware. IDLE-TICs represent the hardware/software interface when there is no exchange of information between hardware and software.

At every time frame of the unrolling, exactly one TIC is instantiated, modeling the hardware/software interactions that can occur at that time point. A DATA-TIC is instantiated if there is data exchange at the given time frame, otherwise an IDLE-TIC is used. The signals of a TIC instance are connected to the corresponding copy of the uncore hardware in the unrolling (cf. Fig. 4.8).

Since an instruction can be executed at different time points, depending on the particular path executed by the software, a single input/output instruction cell can connect to different DATA-TICs. In the same way, a single DATA-TIC can connect to different input/output instruction cells since for different execution paths different input/output instruction cells can be active at the time point defined by the related DATA-TIC. Section 4.5.2.2 describes how input/output instruction cells connect to DATA-TICs by employing resolution blocks. Contrary to DATA-TICs, IDLE-TICs do not require connection with any instruction cell of the program netlist.

For example, in Fig. 4.8 the instruction cells i_1, i_4, i_6 and i_8 (marked in blue) correspond to input/output instructions and therefore connect to DATA-TICs. In the example of Fig. 4.8, the white instruction cells do not exchange data with the hardware and therefore IDLE-TICs are instantiated in the interface model for representing them. Note that there is no connection between IDLE-TICs and the white instruction cells. In the case of interactions with portions of the surrounding hardware that are irrelevant to the part of the system under verification the behavior of the interface is also modeled using only IDLE-TICs.

Figure 4.9 shows an example of how a STORE instruction writing to the hardware is modeled using different TICs. (A LOAD instruction reading values from the hardware can be represented in a similar way.) In this example, the processor has a

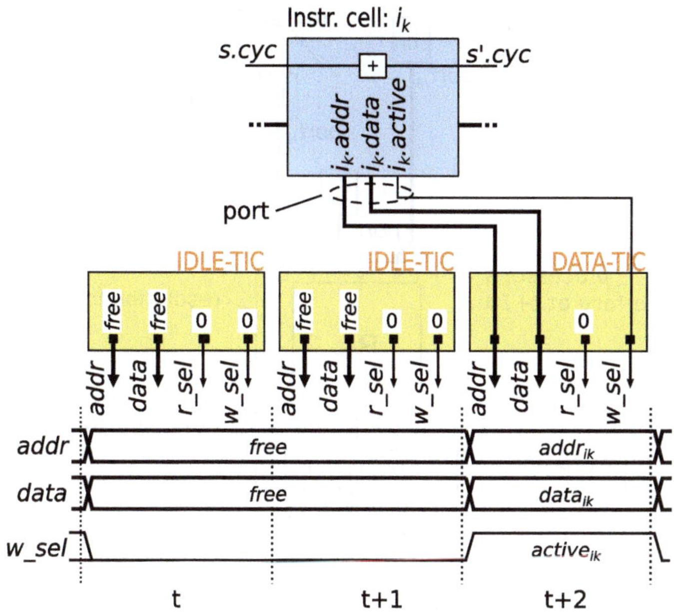

Fig. 4.9 Example of TICs for a non-pipelined multi-cycle architecture

non-pipelined three-phase multi-cycle architecture. A write is executed in the third clock cycle. For each ISA instruction three TIC instances are needed. As shown in Fig. 4.9, the first two clock cycles are represented by IDLE-TICs since during these clock cycles no data is exchanged. Also, there is no connection of these IDLE-TICs with the program netlist. Data transfer happens in the third clock cycle which is modeled by a DATA-TIC. This TIC connects the hardware signals of the third time frame with the port of the STORE instruction cell.

The example also shows how non-determinism is used in modeling the processor's input/output interface through TICs: The control signal *w_sel* specifies the validity of the output data *data*. At time points where *w_sel* is 0 the *data* signal is left undetermined (modeled by an unconstrained "free" input). In this way, details of the processor's implementation which are not relevant to the model are abstracted away.

4.5.2.2 Input/Output Resolution Logic

If two or more input/output instructions can possibly access the interface of the processor at the same clock cycle then extra control logic is needed to resolve which instruction actually drives the interface signals depending on the execution path taken in the software. Figure 4.10 shows an example where instructions i_4, i_6 and i_8 (from Fig. 4.8) can write to the hardware at the same clock cycle. A resolution logic block

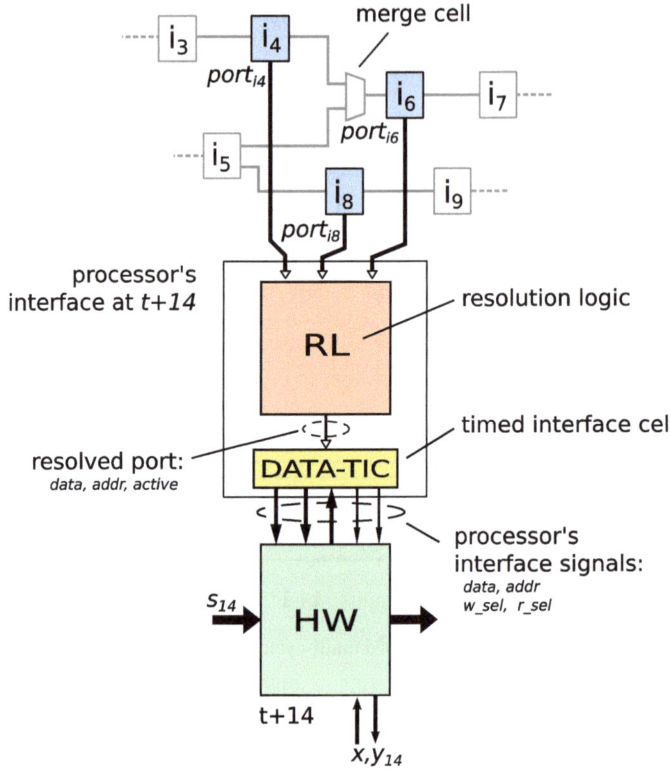

Fig. 4.10 Example of unrolled HW/SW model containing a resolution block

(RL) is instantiated that decides which of these input/output instructions cells drive the interface signals.

The resolution logic takes as input all port signals of the involved input/output instruction cells and decides which of them are connected to the corresponding DATA-TIC. In general, this decision depends on the active signals values (which in turn depend on the inputs of the program) and on whether accesses are performed at the given clock cycle. Since the program netlist is a DAG with paths re-converging at merge nodes, individual instruction cells in the model may belong to different execution paths. Consequently, a single instruction cell (for example i_6 of Fig. 4.10) can access the interface at different time points. To resolve this, we add a time variable called *cyc* to the program state determining the clock cycles at which instructions are executed. A counter incrementing the number of clock cycles is added to all instruction cells of the model updating the value of *cyc* (see the instruction cell in Fig. 4.9).

If at clock cycle k the set of input/output instruction cells $W_k = \{i_1, i_2, ..., i_m\}$ can access the interface then the resolution logic for the output data value ($data(k)$) is defined as follows.

$data(k) :=$

$$
\begin{aligned}
&\quad\textbf{\textit{if}}\ (i_1.active\ \ \textbf{\textit{and}}\ \ (i_1.cyc = k))\ \textbf{\textit{then}}\ i_1.data\\
&\textbf{\textit{else if}}\ (i_2.active\ \ \textbf{\textit{and}}\ \ (i_2.cyc = k))\ \textbf{\textit{then}}\ i_2.data\\
&\quad\ldots\\
&\textbf{\textit{else if}}\ (i_m.active\ \ \textbf{\textit{and}}\ \ (i_m.cyc = k))\ \textbf{\textit{then}}\ i_m.data\\
&\textbf{\textit{else}}\ free
\end{aligned}
$$

This logic describes a chain of multiplexers in which $i_l.data$, $i_l.addr$ and $i_l.active$ are the port signals of the instruction cell i_l. For the other port signals, $addr_k$ and $active_k$, a similar multiplexer chain can be constructed. Input ports are handled in a similar same way.

In order to find the set of instructions W_k for $k = 0, ..., n - 1$ (with n being the total number of clock cycles of the unrolling) an algorithm that propagates the execution times for each individual instruction cell through the execution graph is employed. The algorithm takes as inputs the execution graph, the address space reached by the program and timing information of the processor's architecture which helps to predict statically when instructions can be executed in time. Note that this process is similar to the one explained in Sect. 4.4.1 with the difference that here the resulting model is cycle-accurate. The first two inputs to the algorithm were already computed previously when generating the program netlist.

The kind of timing information needed as input depends on the particular processor architecture. In the case of a non-pipelined multi-cycle architecture, for example, the number of clock cycles per instruction is needed as well as the timing specific to phases where input/output access (e.g., memory address stage) takes place. For pipelined architectures it is additionally required to provide a description of stall scenarios. In the following we describe the algorithm for the case of a non-pipelined multi-cycle architecture.

The algorithm traverses the execution graph which has been topologically sorted. Associated to every EXG node q there is a set of possible execution times T_q at which the instruction at the node can be executed. When visiting a new EXG node q, the set T_q is computed by taking the execution times of all predecessors and adding to them the delay of the current instruction. (In the case of a non-pipelined architecture this delay is a constant.) Since EXG nodes are sorted topologically, it is ensured that when a given node is processed the execution times of all its predecessors are known. The algorithm ends when all EXG nodes have been processed.

We consider again the example of Fig. 4.8 and assume a three-phase multi-cycle architecture. The algorithm begins with computing $T_{i_0} = \{0\}$, then $T_{i_1} = \{3\}$, and so on. When i_6 is processed, it is already known that $T_{i_4} = \{12\}$ (this is ensured since the traversal is done respecting the topological order of the EXG), $T_{i_5} = \{9\}$ and therefore $T_{i_6} = \{12, 15\}$ is computed. The algorithm can then further continue to compute $T_{i_7} = \{15, 18\}$. Execution times for the remaining nodes are computed in the same way. The traversal ends once the possible execution times are calculated for i_10.

After the algorithm finishes, in a second step, the information of the address space reached by input/output instructions is analyzed in order to detect input/output instructions cells that interact with the hardware. All other instruction cells are marked as *IDLE* and therefore are represented in the time interface model by a number of IDLE-TICs instances corresponding to the clock cycles spent in these instructions.

In our example, it can be seen that during $t + 12$ and $t + 14$ instruction cells i_4, i_6, i_8 can be executed. More specifically, since in the last phase the data is written to the hardware it can be determined that $W_{14} = \{i_4, i_6, i_8\}$, as shown in Fig. 4.10. All other time points of the unrolling can be handled in the same way.

Note that this approach significantly reduces the amount of logic needed since resolution logic blocks are instantiated only at points when they are in fact needed, and each resolution block takes into account only relevant input/output instruction cells.

4.5.3 Experimental Results

We consider an application domain as described in Sect. 4.3 where high predictability of the processor's timing behavior allows for a firmware-based design style, in which the processor is directly integrated into the hardware without the use of a standard bus interface.

The verification platform *Formal Checker Kaiserslautern* (FCK) was extended with the algorithms and modeling elements presented in Sect. 4.5.2. For program netlist generation, FCK takes as input the machine code of the firmware and the instruction cells modeling the core hardware. In order to generate the timed interface model, FCK requires an RTL description of the TICs of the processor together with the specific timing information for predicting instruction execution times. Finally an RTL description of the uncore HW is also needed. With this information FCK fully automatically generates the hardware/software model for verification.

The FCK backend generates the combined hardware/software model such that it can be used as input to a standard property checker. In the following experiments the commercial property checker OneSpin 360 DV [15] was used. All experiments were conducted on an Intel Xenon running at 2.83 GHz with 32 GB of main memory.

Two different case studies were conducted employing the PicoBlaze processor from Xilinx. Properties were written (using OneSpin's property language ITL) that specify global behavior of the designs, e.g., complete transactions. For both case studies the approach of [19] was followed extended by the techniques of Sect. 4.5 to generate program netlists representing the firmware behavior. In all experiments the approach presented in this chapter was compared with the classical hardware bounded model checking (BMC) technique as described in Sect. 4.2. The authors are not aware of any other tools or methods reported in the literature that could be applied in this context.

The first case study is a firmware-based implementation of a *slave interface* for the *Flexible Peripheral Interconnect* (FPI) bus protocol. The FPI bus is a pipelined SoC bus developed by Infineon. The slave serves to connect a peripheral device

Table 4.4 Characteristics of the designs and models

Design	LoC		PN	DATA-	HW	state
	SW	HW	size	TICs	inputs	vars.
FPI slave	172	2908	380	188	113	1149
Control unit	253	2734	474	171	18	1148

LoC HW: lines of HDL code, LoC SW: lines of assembly code,
PN size: number of instruction cells

Table 4.5 Property checking results

Property	PN-based			BMC-style	
	length (cycles)	time (min.)	mem. (GB)	time (min.)	mem. (GB)
slave_write	461	01m21s	1.93	45m02s	19.11
slave_read	460	01m50s	1.82	43m34s	20.78
trans_ok	729	04m02s	1.62	TO[a]	23.40[b]
trans_valid	700	03m04s	1.46	TO[a]	21.05[b]

[a]*time-out = 24h,* [b]*memory usage at time-out*

with the FPI bus. It interacts synchronously with the bus and asynchronously with the peripheral. In the system, the surrounding hardware (wrapper RTL in Fig. 4.5) captures the signals from the bus when a request occurs. Subsequently, the firmware checks the incoming request and informs the peripheral about it. Once the answer from the peripheral arrives, the firmware finishes the transaction by setting the correct values on the wrapper RTL. Table 4.4 summarizes the characteristics of the design and the models. Proofs were conducted for two different safety properties *slave_read* and *slave_write*, specifying a read and a write transaction, respectively, after a system reset sequence. Both properties describe control and data values as specified in the FPI bus documentation. Table 4.5 presents the results obtained for proving both properties.

The second case study is a control unit resembling typical non-mainline functionality implemented with the use of firmware. The control unit interacts on one side with a master SoC module (e.g., a processor or some other hardware module) which sends commands to specific hardware devices through the control unit. Addresses of the destinations (hardware units) are also sent to the control unit by the same master SoC module. The control unit receives command and destination information, analyzes its validity (checking parity) and then synchronously sends the command to the corresponding hardware devices. Each of the hardware devices runs an independent finite state machine recognizing whether commands are actually intended for it or not. In case of a match, the command is latched by the hardware unit and a control signal is activated to indicate the arrival of the valid command. At the end, the control unit informs the master SoC module whether the transaction has been completed successfully or not. Table 4.4 summarizes the characteristics of the design and the resulting models. Two safety properties *trans_ok* and *trans_valid* were proved. The first property specifies that (after a reset sequence was applied) if the data obtained from the master SoC module is valid then at the end of the transaction the corresponding commands and control signals are activated in the correct hardware unit. The second property specifies that the correct finish condition is sent to the master SoC module depending on the validity of the command and destination values. Table 4.5 contains the results obtained for proving both properties.

All properties were finally proven correct after a number of system bugs had been identified by the method and were corrected. For instance, in the FPI slave interface there was a situation in which after the reset, the slave could be selected before the system was properly initialized by the firmware and therefore wrong values of the FPI control signals were issued by the interface. Also the firmware of the control unit contained a bug due to a wrong sequence of reading operations. For a certain execution scenario of the firmware it was possible that the acknowledge was read after the command value. This could trigger invalid read commands by the control unit.

As can been observed, the proposed technique outperforms a straightforward BMC approach in all cases. The properties for the control unit turn out to be more difficult for the property checker (i.e., for the SAT engine) than the properties for the FPI bus. An explanation for this is that the control signals depend on the parity computation performed by the firmware which increases the computational challenges for the solver.

In both case studies, the firmware was required to react as fast as possible so as to cause only a minimum number of wait states. Therefore, the run times for complete transactions, in both cases, are short, resulting in program netlists of small size. Their generation required less than 1 min in all cases. Taking into account that much larger program netlists have already been generated successfully in [19, 20], there is promise that also significantly larger designs integrating firmware as described in Sect. 4.5.1 can be handled.

4.6 Conclusion

This chapter described hardware-dependent modeling of software based on program netlists. Two application scenarios for program netlists have been addressed, namely, speed-independent communication of the processor with its periphery and cycle-accurate integration of firmware into an SoC module.

The proposed methods benefit from information provided by program netlists which facilitates modeling of the hardware/software interface for different time granularities. In particular, information about the memory address spaces accessed by the software is used to identify possible hardware/software interactions. Additionally, explicit control flow representation is employed for determining temporal information about those hardware/software interactions.

For speed-independent communication schemes, a time-abstract model of the hardware/software interface has been proposed for checking equivalence of two reactive programs. For cycle-accurate integration, a time interface model is constructed and integrated to the model for performing hardware/software co-verification.

As shown by the experiments, in both scenarios low-level software of realistic complexity can be handled by the proposed approaches. This encourages further investigations also in other application domains such as testing and formal safety analysis of hardware/software systems.

References

1. T. Aitch, Aquarius: a pipelined RISC CPU (2003)
2. T. Arons, E. Elster, L. Fix, S. Mador-Haim, M. Mishaeli, J. Shalev, E. Singerman, A. Tiemeyer, M.Y. Vardi, L.D. Zuck, Formal verification of backward compatibility of microcode, in *Proceedings of the 17th International Conference on Computer Aided Verification, CAV'05* (Springer, Heidelberg, 2005), pp. 185–198
3. T. Arons, E. Elster, S. Ozer, J. Shalev, E. Singerman, Efficient symbolic simulation of low level software, in *Design, Automation and Test in Europe, DATE '08* (2008), pp. 825–830
4. B. Bao, C. Villarraga, B. Schmidt, D. Stoffel, W. Kunz, A new property language for the specification of hardware-dependent embedded system software, in *Proceedings of Forum on Specification and Design Languages (FDL)*, Munich, Germany, Oct 2014. (accepted for publication)
5. C. Bartsch, C. Villarraga, B. Schmidt, D. Stoffel, W. Kunz, Efficient SAT/simulation-based model generation for low-level embedded software, in *17. GI/ITG/GMM Workshop Methoden und Beschreibungssprachen zur Modellierung und Verifikation von Schaltungen und Systemen (MBMV)* (2014), pp. 147–157
6. A. Biere, A. Cimatti, E. Clarke, Y. Zhu, Symbolic model checking without BDDs, in *Proceedings of the 5th International Conference on Tools and Algorithms for Construction and Analysis of Systems, TACAS '99* (Springer, London, 1999), pp. 193–207
7. C. Cadar, K. Sen, Symbolic execution for software testing: three decades later. Commun. ACM **56**(2), 82–90 (2013)
8. D. Currie, X. Feng, M. Fujita, A.J. Hu, M. Kwan, S. Rajan, Embedded software verification using symbolic execution and uninterpreted functions. Int. J. Parallel Program. **34**(1), 61–91 (2006)
9. D.W. Ecker, V. Esen, T. Steininger, Memory models for the formal verification of assembler code using bounded model checking, in *Proceedings of IEEE International Symposium on Object-Oriented Real-Time Distributed Computing* (2004), pp. 129 –135
10. N.Eén, N.Sörensson. An extensible SAT solver. in *SAT* (2003), pp. 502–518
11. A. Hazra, R. Mukherjee, P. Dasgupta, A. Pal, K. Harer, A. Banerjee, S. Mukherjee, Powertructor: An integrated tool flow for formal verification and coverage of architectural power intent. IEEE Trans. Comput. Aided Des. Integr. Circ. Syst. **32**(11), 1801–1813 (2013)
12. A. Horn, M. Tautschnig, C. Val, L. Liang, T. Melham, J. Grundy, D. Kroening, Formal co-validation of low-level hardware/software interfaces. Formal Methods Comput. Aided Des. (FMCAD) **2013**, 121–128 (2013)
13. J. Koestersm, A. Goryachev, Verification of non-mainline functions in today's processor chips. in *Proceedings International Design Automation Conference (DAC), DAC'14* (ACM, New York, 2014), pp. 1:1–1:3
14. M.D. Nguyen, M. Thalmaier, M. Wedler, J. Bormann, D. Stoffel, W. Kunz, Unbounded protocol compliance verification using interval property checking with invariants. IEEE Trans. Comput. Aided Des **27**(11), 2068–2082 (2008)
15. Onespin Solutions GmbH. Germany. OneSpin 360MV. http://www.onespin-solutions.com

16. A. Pnueli, M. Siegel, E. Singerman, Translation validation, in *Proceedings of the 4th International Conference on Tools and Algorithms for Construction and Analysis of Systems, TACAS '98* (Springer, London, 1998), pp. 151–166
17. C.S. Păsăreanu, W. Visser, A survey of new trends in symbolic execution for software testing and analysis. Int. J. Softw. Tools Technol. Transf. **11**(4), 339–353 (2009)
18. B. Schlich, Model checking of software for microcontrollers. ACM Trans. Embed. Comput. Syst. **9**(4), 36:1–36:27 (2010)
19. B. Schmidt, C. Villarraga, T. Fehmel, J. Bormann, M. Wedler, M. Nguyen, D. Stoffel, W. Kunz, A new formal verification approach for hardware-dependent embedded system software, *IPSJ Transactions on System LSI Design Methodology (Special Issue on ASPDAC-2013)*, vol. 6 (2013), pp. 135–145
20. C. Villarraga, B. Schmidt, C. Bartsch, J. Bormann, D. Stoffel, W. Kunz, An equivalence checker for hardware-dependent software, in *11 ACM-IEEE International Conference on Formal Methods and Models for Codesign* (2013), pp. 119–128
21. M. Wedler, E. Cabrill, S. Graham, L. Patrick, Using formal verfication for HW/SW co-verification of an FPGA IP core. Xcell Journal (Xilinx, Inc.) **0**, 56–61 (2012)

Author Biographies

Carlos Villarraga received his undergraduate degrees in electrical and electronics engineering in 2002 and master degree in microelectronics and computer engineering in 2006 from Universidad de los Andes, Bogota, Colombia. In 2016, he obtained the Ph.D. degree at the Electrical & Computer Engineering at Technische Universität Kaiserslautern. Currently he holds a post-doctoral position at the same university. His current research interest includes formal verification of embedded hardware/software systems.

Dominik Stoffel received the Dipl.-Ing. degree from the University of Karlsruhe, Germany, in 1992, and the Ph.D. degree from the Goethe University Frankfurt, Germany, in 1999. From 1993 to 1994, he was a Research and Development Engineer with Mercedes-Benz, Stuttgart, Germany, in the development of testing methodology for automotive electronics. From 1994 to 1998, he was with the Max-Planck Fault-Tolerant Computing Group, Potsdam, Germany. From 1998 to 2001, he was a research scientist at the University of Frankfurt/Main. Since 2001, he is a research scientist at the Department of Electrical & Computer Engineering at Technische Universität Kaiserslautern, where he became a professor in 2012. He conducts research in the area of design and verification of embedded systems and systems-on-chip. He has a special interest in methodologies based on formal verification of hardware and low-level software.

Wolfgang Kunz received the Dipl.-Ing. degree in Electrical Engineering from the University of Karlsruhe, Germany, in 1989 and the Dr.-Ing. degree in Electrical Engineering from the University of Hannover, Germany, in 1992. From 1993 to 1998, he was with Max Planck Society, Fault-Tolerant Computing Group at the University of Potsdam, Germany. From 1998 to 2001, he was a professor of Computer Science at the Goethe University Frankfurt, Germany. Since 2001, he is a professor at the Department of Electrical & Computer Engineering at Technische Universität Kaiserslautern. The focus of his current research activities is on design and verification of Embedded Systems and Systems-on-Chip (SoC). Research topics include formal verification of hardware and hardware-dependent software, system-level design flows, safety analysis and design for power closure.

Chapter 5
Formal Verification—The Industrial Perspective

Raik Brinkmann and Dave Kelf

5.1 Introduction

Although formal technology has been applied to verification for many years, its widespread adoption has been limited. However, recently formal verification has become a key component within modern design flows. So what has changed? Several trends have driven adoption of formal techniques (see e.g. [1–7]).

Clearly, more powerful underlying technology is a key driver for formal adoption, allowing the expansion of the scope of application. Such advancements include improvements in basic algorithms, such as Boolean Satisfiability (SAT), as well as more powerful SAT-based model checking techniques, such as Bounded Model Checking (BMC) and IC3. While these advancements are mainly driven by academic research, they are quickly adopted by the industry, leveraging them into commercial tools. While formal adoption is increasing, the demand for further improvements on this level is relentless.

A key problem with formal adoption in the past was the inaccessibility of this approach to a wider audience because of a plethora of different and complex modeling concepts and mechanisms. While still complex, the availability of standard design and verification languages that work across many tools in the design flow such as VHDL, SystemVerilog, PSL (Property Specification Language), and SVA (System Verilog Assertions) facilitates formal adoption. It allows the more complex concepts to be hidden underneath a common framework that is more accessible to design and verification engineers. Using these standards on an industrial scale further lowers the barrier of entry for formal, as it makes the creation of automated tools a worthwhile

R. Brinkmann (✉) · D. Kelf (✉)
OneSpin Solutions, Munich, Germany
e-mail: raik.brinkmann@onespin-solutions.com

D. Kelf
e-mail: dave.kelf@onespin-solutions.com

R. Drechsler (ed.), *Formal System Verification*, DOI 10.1007/978-3-319-57685-5_5

endeavor. In Sect. 5.2 we will look at some automated formal techniques, ranging from early design analysis, to SoC integration and implementation.

While automation is a key driver for the widespread adoption of formal, automation can't yet solve the functional verification problem if the specification is informal. In this case, the standard practice is to break down the specification into ever-smaller functional components, and to manually capture these components using formal properties. These formal properties are then verified against the implementation. This constitutes the classic model checking approach, referred to as formal Assertion-Based Verification (ABV). Driven by advanced formal algorithms, the scope of model checking has steadily increased over the past two decades, making more and more designs accessible to formal ABV. With increased adoption, questions of solution convergence, predictability, and quantification of verification become more important. Another important question relates to the adequacy of standard property languages for manual and automated inspection, regarding completeness and consistency. In Sect. 5.3 we will give an overview of functional formal verification practices and how these challenges can be addressed.

New integrated circuit applications create additional design and verification challenges. In particular, the convergence of advanced data processing, wireless connectivity, and re-configurability in high-reliability applications such as advanced driver assistance systems (ADAS), wireless sensor networks, smart factories, and other IoE domains has an impact on how systems are designed and verified. For example, higher levels of abstraction, and the need to ensure functional safety and hardware security add to verification complexity. In these areas new verification approaches and technologies are needed in order to keep up with demand. In Sect. 5.4 we will investigate some recent industry trends and verification solutions that address them.

5.2 Automating Design Verification with Formal

In order to increase design productivity and agility while facilitating designer creativity, early detection of design issues, the automation of recurring verification tasks, and prevention of functional regressions throughout the design refinement process are key components. We will look at examples of automated formal usage throughout the design process, from RTL design entry, to the block integration and implementation phases.

5.2.1 Design Inspection

The early elimination of bugs in an IC development process saves time and resources. This places pressure on component designers to perform more verification. However, given traditional simulation-based verification techniques, the ultimate result of this trend is designers spending more time creating stimulus and getting involved with overall verification, and less on creative design.

The agile software development movement has proposed a more interactive development methodology, in an attempt to restore creativity and reduce administrative boundaries. These ideas translate well into the IC hardware design process in general, but require new thinking in terms of verification techniques.

Today, design teams employ varying solutions to get their code right. These include manual inspection, linting tools to highlight possible errors and coding issues, and both interactive and batch simulation. While simulation allows design operation to be observed, it requires a fair amount of setup and testbench creation. Linting reduces setup overhead, and performs a useful but relatively "dumb" check that highlights possible errors, however it requires sifting through lots of data to identify real issues.

Successful adoption of agile principles in hardware design requires the automation of verification, and regular review of design artifacts is key. Here, formal technology can be of great value. We will now look at three examples, property generation, operational inspection, and activation analysis.

5.2.1.1 Property Generation

A broad range of bugs may be introduced into design blocks either by coding accidents or a misunderstanding of expected Verilog or VHDL semantic operation. Some coding problems can be identified through the use of Linting, while most will require simulation. Many bugs will only be detected late in the process, almost certainly leading to a large expenditure of debug resources to repair. In particular, the VHDL and Verilog/SystemVerilog standards allow for a range of coding scenarios that can lead to bugs. For example, consider an array with n elements, accessed through an index that can potentially have values beyond the $[0, n - 1]$ range, e.g., a 6-element array indexed by a 3-bit signal. In the case of VHDL, the standard prescribes that a simulator must generate a Fatal Design Error (FDE) and stop running if during a simulation the array is indexed out of its bounds. Of course, this requires stimulus to be executed to allow its discovery during RTL simulation. In the case of Verilog the standard prescribes the generation of X states in this scenario, requiring the problem to be stimulated and the cause of the X state propagated to testbench checks. In both cases a synthesized netlist will have a defined behavior that does not match the RTL simulation semantics, which can lead to the issue being hidden at the netlist level. This issue is easy to accidently create, and can be hard to detect during simulation, or will often be buried as one of many "out-of-bound access" warnings in a linting log file.

Instead of simply analyzing the code structure with the risk of many false positive errors, formal tools can generate formal properties that catch these unwanted scenarios by inspecting the actual operation of the code, and automatically check if these scenarios are reachable. Unlike linting, a formal tool eliminates log files and many false warnings and produces a simulation trace that can be debugged. As compared

```
reg [2:0] i;
reg [5:0] array;

always @(posedge clk)
for(i=0 ; i<=5; i=i+1)
  array[i] <= 1;
```

Lint: This code "might" result in out of bounds access

Inspect: Design operation does not result in an out-of-bounds access

Fig. 5.1 Array out-of-bounds inspection versus linting

to the simulation, the formal tool will consider ALL possible operational details automatically without the need for stimulus. For example, if the design is capable of an out of bounds access, the error will be flagged and the code driver creating that mechanism will be highlighted without relying on the user to provide simulation test vectors that trigger this scenario (see Fig. 5.1).

Another classic scenario that often produces a Simulation-Synthesis Mismatch (SSM) is the misuse, in Verilog, of the Synopsys "Full-Case" and "Parallel Case" pragmas. These pragmas instruct the synthesis tool that certain logic optimizations can be safely made in order to improve area and timing. However, they have no effect on the RTL simulation semantics and if the designer has made a mistake, for example using the Parallel-Case pragma in a case statement with branches that are not mutually exclusive, the behavior of the synthesized netlist may potentially not match the RTL semantics.

Two keywords have been introduced in SystemVerilog, "unique" and "priority", to mitigate this problem. The use of the keyword unique is equivalent to using both the Parallel-Case and Full-Case pragmas. The use of the keyword priority is similar to using the Full-Case pragma. The advantage is that the standard prescribes that a simulator must issue a warning if a scenario is hit where the case statement does not respect the unique or priority semantics. This will help in flagging a problem, but once again the scenario must be uncovered by the simulation stimulus. To cover this scenario, the formal tool will synthesize assertions whenever this style of case statement is used, for both the SystemVerilog key words and RTL synthesis pragmas. It will then exhaustively check that the code has been correctly written and that no simulation-synthesis mismatches can occur (Fig. 5.2).

There are in fact many of these cases, where classic RTL errors may be avoided by the judicious use of formal apps. The creation of design intent assertions from RTL code is a good example of automated formal verification. Automatic formal design inspection can check for a number of design issues, examining code operation for real problems, rather than code syntax for possible issues. It saves hours of manual analysis while detecting tougher bugs. Integrated into nightly regressions, it

```
case (1'b1) // synopsys parallel_case
  i1: o1 = 1'b1;
  i2: o2 = 1'b1;
  default :
    begin
      o1 = 1'b0;
      o2 = 1'b0;
    end
```

Can **i1** and **i2** be active simultaneously?

Fig. 5.2 Parallel-case example

automatically and reliably shortens the time between the introduction of issues and detection.

5.2.1.2 Operational Inspection

The early elimination of bugs in an IC development process saves time and energy. This places pressure on component designers to perform more verification. However, given traditional simulation-based verification techniques, the ultimate result of this trend is designers spending more time creating stimulus and getting involved with overall verification, and less on creative design.

In contrast, the Agile software development movement has proposed a more interactive development methodology in an attempt to restore creativity and reduce administrative boundaries. Agile suggests an interactive model where designers create, quickly test, and integrate small, valuable code sections. These ideas translate well into the IC hardware design process in general, but require new thinking in terms of verification techniques.

What is required is a verification solution with a low setup and usage overhead that demonstrates design operation without even having to consider stimulus creation, and automatically track real design issues in depth. Using formal technology, it is possible to generate witnesses from properties describing some desired design operation. These very simple properties (e.g., assume reset goes low then high) are typically specified using some graphical or textual input method that is much simpler than standard property specification languages. Instead of directly writing complex sequential properties, a tool can generate properties from these alternative mechanisms, eliminating the need for the user to learn a formal property language, thus increasing design productivity. A multitude of desired design operations is typically kept as a list of scenarios, each reflecting a particular operational aspect. A key advantage of these scenarios is that they are more robust in the face of implementation changes in the design than traditional directed simulation tests. This is because the actual bit wiggling of the inputs driving the design can be regenerated by the formal tool automatically.

In this manner, one can observe design operation with no or little stimulus, which allows a designer to see what is going on in the code quickly and effortlessly. In this way test creation is automated, and can even improve later simulation if necessary.

5.2.1.3 Activation Analysis

Deadlocks in finite-state machines, unreachable code, and stuck signals are major concerns when developing and integrating RTL components. Usually they point to functional errors or misinterpreted integration conditions that render certain design functionalities inactive. These might include miscalculated branching conditions, incorrect wiring, or interlocking conditions. These errors cannot reliably be detected using simulation, which can only demonstrate the activation of certain design func-

```
case (state)
   2'b00: nstate = 2'b01;
   2'b01: nstate = 2'b11;
   2'b10: nstate = 2'b00;
   2'b11: if (ack)
              nstate = 2'b10;
          else
              nstate = 2'b11;
endcase
```

Can this branch be covered?

Fig. 5.3 FSM coverage example

tions, but can never prove the opposite case, the inability of such design sections to be activated.

Formal is the only technology that can prove the absence of capability, that is, it is able to show that it is impossible to activate some function. This information is useful both for debugging these conditions, as well as creating coverage waivers for simulation if the function is indeed not relevant in the context of the design (e.g., design sections used for debug and not in normal operation). If it is possible to activate a function, a formal approach will find out and construct a sequence of stimulus that demonstrates the activation. The generated trace can then be used for the interactive analysis of design behavior, or the creation of a simulation test bench.

Typical activation checks automated by formal are code reachability checks, finite-state machine checks, and signal toggle checks:

1. Code Reachability Checks target each branch of input source code, and check whether the branching condition can be activated. Both if-then-else and switching conditions (case) are considered, and the default behavior analyzed. This is particularly helpful to avoid don't-care conditions, via explicit X assignments, that could lead to simulation-synthesis mismatches and other hazards.
2. Finite-State Machine (FSM) Checks verify that each transition of an FSM can be performed, that no code deadlocks exist, and that the FSM correctly returns to an initial state on reset (Fig. 5.3). FSMs can be automatically identified in the input source code. The checks can be performed across the entire design function, and can include other FSMs that might interact with the machine under consideration. This is particularly helpful for finding unwanted interlocking conditions between multiple FSMs.
3. Toggle Checks analyze the switching capability of signals in the design. The checks depend on the type of the signal. For example, in case of a Boolean signal, it is useful to check if it can transition from 0 to 1, and from 1 to 0. For other signal types such as numeric and enumerated signal types other checks are appropriate.

Typically, activation checks are used during early design stages to analyze newly written design functions. If a function cannot be activated, it suggests incorrectly coded execution and branching conditions. During functional verification, activation checks can be helpful to analyze simulation coverage holes. Additionally, during IP integration, if design functions can no longer be activated after integrating a block, this suggests a problem with the assumptions regarding the integration conditions, wiring, etc.

5.2.2 *IP Integration Verification*

Another aspect that drives formal adoption is increased use of System-on-Chip (SoC) methodologies. A SoC methodology drives designers to create individual components, which are then integrated onto an SoC platform, which in turn is verified continuously in lengthy regression simulation runs. This approach also produces a higher demand for pre-verified functional blocks of intellectual property (IP), allowing the integration of pre-verified components into an SoC. A modern SoC will also contain a number of processors and specific hardware and IP blocks, which are connected through on-chip bus interfaces. Almost every block in an electronic design contains a series of registers. These registers are used to configure, control, and monitor the operation of the block, often loaded or read from a system processor utilizing a related driver software function.

This IP integration methodology has created a whole new class of automated formal tools, or formal "apps" to verify SoC connectivity, on-chip-protocol compliance of IP blocks, register map validation at the hardware–software interface, data transport issues, the propagation of uninitialized states and others. It is critical to eliminate as many bugs as possible prior to components being released into the regression simulation run, as tracking down bugs at this point is extremely expensive. The key to success is the automation of recurring verification tasks based on standards, such as SystemVerilog and IP-XACT. Formal Apps are ideal for this purpose.

5.2.2.1 On-Chip Protocols

A number of standard on-chip bus protocols, such as OCP, AHB, AXI, and other standards are in use today in modern SoCs. These protocols often involve complex sequences of transactions and signal manipulations. When integrating IP blocks, especially those obtained from third party sources, it is critical that inter-block communication mechanisms work properly under all conditions. Many device issues are rooted in the failures of these mechanisms, and the resulting bugs can be hard to track down and fix.

The common method to test on-chip protocols is through simulation. However, to fully exercise an interface, large stimulus files must be created. Ensuring that all the unique combinations of transactions are fully exposed requires hours of simulation time and is hard to achieve given the many operational permutations. This is an ideal application for a formal-based solution, which can test all possible communication transactions in a specific protocol without the need to specify every single one of them.

Pre-designed formal property sets in conjunction with formal assertion-based verification, can be applied in a fully automated manner to a bus interface and will perform a thorough, exhaustive test of this interface with no requirement for the creation of stimulus. Instead of writing a new set of properties, pre-designed property sets are reused and simply instantiated on a specific design under verification. The

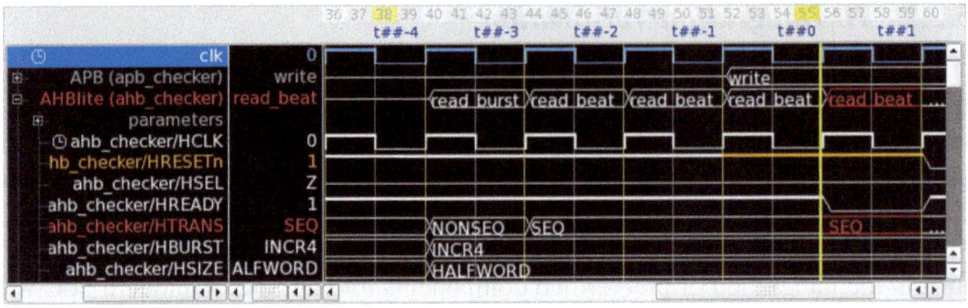

Fig. 5.4 AHB protocol debug display

assertions can expose the existence of specific violations, or formally prove that the interface is performing correctly. The formal tool is also able to trace the operation of the protocol on a particular design example, providing a visual simulation waveform-like display that may be used to inspect behavior (Fig. 5.4)

In order to maximize reuse and interoperability, formal verification IP is usually created using standard assertion languages like SystemVerilog Assertions (SVA). However, many industry applications only provide a readable interface to their IP, and encrypt the interior to protect their property, which limits the reuse within the design flow.

5.2.2.2 Register Map Validation at the Software Interface

IP blocks communicate to software through a range of software accessible registers. Some blocks may contain hundreds of registers, with a nested implementation to handle addressing. The register addresses are defined by a memory or register map, a common document used throughout hardware, software, system/IP integration, and verification engineering to ensure consistency. This map might form part of a specification in the case of the IP block coming from a third party.

A very common source of errors in a system is a mismatch between some aspect of a register and its specification. Although this would seem to be a relatively trivial issue, it can cause many wasted hours spent in debug, particularly because the problem might not manifest itself until the block software driver is operating with application software, at which point multiple engineering functions are involved with disassociated knowledge of the system. Carefully reading a modern memory map specification to ensure system consistency takes many hours, and is extremely error prone given the typically large number of registers.

Today, register operation is often checked using simulation, which is inadequate for this purpose as it is entirely dependent on the quality of stimulus, which leads to missed issues. An exhaustive analysis of register behavior is required where all register interaction, operation, and consistency is tested completely. In simulation a series of read and write interactions would require testing for each register, as well as ensuring that registers behave independently of each other, clearly requiring a po-

```
<spirit:memoryMap>
    <spirit:name>System_APB</spirit:name>
    <spirit:addressBlock>
      <spirit:name>vcore</spirit:name>
      <spirit:baseAddress>0x8000</spirit:baseAddress>
      <spirit:range>0x2c</spirit:range>
      <spirit:width>16</spirit:width>
      <spirit:register>
        <spirit:name>debug</spirit:name>
        <spirit:description>debug register</spirit:description>
        <spirit:addressOffset>
          0x0
        </spirit:addressOffset>
        <spirit:size>16</spirit:size>
        ...
```

Fig. 5.5 IP-XACT Register map segment

tentially large stimulus file. An exhaustive check may only be reliably accomplished with a formal solution, which eliminates the need to create a stimulus file all together.

A register specification is often provided for a system using a standard such as the IEEE-1685 IP-XACT format, an XML meta-data representation designed to ease the integration of IP (see Fig. 5.5). This standard is machine-readable and provides an easier way to check for register consistency, providing an ideal input into the formal verification process. From this specification, the formal tool can automatically generate a set of assertions and verify those assertions to a formal verification engine.

A formal register map verification app reads in an RTL description of a block to be tested, together with an IP-XACT file or some other format, that specifies the block's registers, and performs an exhaustive comparison to ensure absolute consistency, by generating automatically a set of assertions and applying those assertions to a formal verification engine. This way, the entire behavior of the register can be formally verified, including its location in the address map, offset location based on a block's base address, register width, and read/write capability, any special extensions provided by the designer, the enumeration of the register, and other attributes included in the specification. For example, in case of nested registers, address mapping should also be verified. Other typical properties of registers include read only, read and reset to one or zero, validation of write only registers, and other typical interactions. Usually, registers that must be updated within specific time periods and behavior on reset may also be tested. Optionally, checks for the absence of registers, or register bits that should be present according to the specification, can be performed.

Some register tools use the IP-XACT format to generate register hardware. In these cases, it has been proven to still be important to verify generated implementations for various situations, including: validating a change to the register map or underlying design, incorrect signal connection, issues with block address bus sizes, extra validation of hierarchical address mapping schemes, and even some RAM tests.

In a typical Systems-on-Chip registers are accesses from the software/processor side through a bus interface, running an on-chip protocol, such as AMBA, AHB, and APB. In order to verify the correct operation of the registers through the bus interface, the use of protocol verification IP (see above) is mandated.

5.2.2.3 SoC Connectivity

The wiring of multiple functional blocks through the hierarchical layers of a design is a tedious and error-prone task. Errors in signal naming conventions, switched positions in port lists, incorrectly matched bus bit positions, and many other fault conditions produce seemingly simple errors that can waste hours of time if undetected until later in the engineering process. On occasions these can be missed completely, resulting in an expensive device re-spin.

The connectivity issue is further compounded by modern interconnect structures (see Fig. 5.6). Complex bus protocols with transaction-level signal propagation, Network-on-Chip (NoC) channels, crossbar and bus bridges, for example, mask signal connections throughout a system, making the verification of correct device connectivity hard and time consuming to establish.

Connection behavior also requires testing, for example, connectivity during reset and power domain switching, a device placed into a test configuration that switches connectivity, and connection options based on control register values. Specialized tools may also be used to automatically establish connections and this process also requires verification. Tracking these issues down using simulation is unreliable as the stimulus used must cover every connection style and behavior in an exhaustive fashion, ensuring linkage through complex interconnect structures throughout the design. The creation of such a test set is also time consuming and error prone.

Formal verification provides an easy way to automate a mechanism to solve this issue. Utilizing a range of machine-readable formats such as connectivity specifications and tables (including spreadsheets), interconnect assertions can be generated that may evaluate connections through the most complex of design structures. Running the design with these assertions through a formal proof engine provides an exhaustive test of all interconnects listed in the table. What appears to be a trivial task gets a bit more complicated if conditional connections, and configurable connections need to be considered, especially when a connection is gated and controlled through register settings, or potentially overridden through device reset or similar events changing the design state.

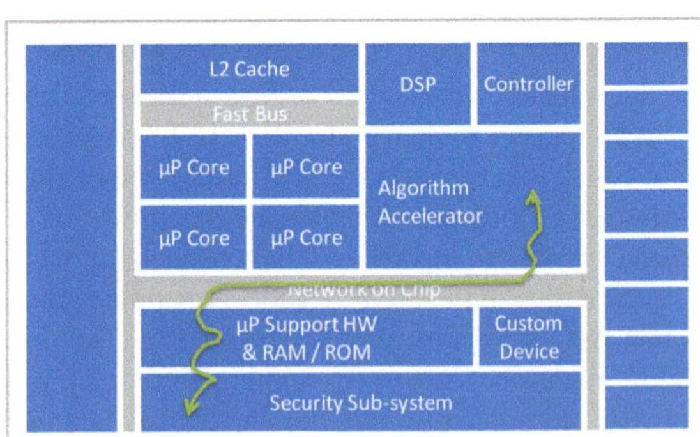

Fig. 5.6 Connectivity checking through bus structures

If an issue is found with a connection, a simulation trace is generated that highlights the problem to allow its inspection. This waveform trace highlights issues regardless of the connection mechanism, allowing for hard to understand connection problems to be observed. In a formal setting, it is possible to quickly ensure that a complete system platform is wired correctly with virtual connections evaluated, eliminating one of the most common reasons for project delays and faulty silicon.

5.2.2.4 Uninitialized (X) State Propagation

Verifying the absence of undefined signal (or X) states, or the conditions that cause them in HDL designs is a critical but often complex operation. However, X states can indicate some of the most dangerous bugs and, as such, analysis to protect against them is a critical component of any verification program.

In general, in both Verilog and VHDL designs, the "X" state is used to depict a signal state that is either undefined or unknown, or has been specifically set in a "don't care" condition. X states can occur due to a variety of reasons, many of which indicate an error condition, and once they do occur can propagate through the design. Often an X state can indicate problems with system or component reset, or gated clocking schemes. On occasion X states may be expected and tolerated, at least for a certain number of clock cycles, and are sometimes used to track other problems. These don't care states are caused generally by explicit X assignments, the testing of a partial reset sequence, or X states being propagated from the input ports. These situations also require verification.

X states are notorious for indicating unusual, corner-case bug conditions and, as such, creating the right stimulus, which might have to control a broad range of signals over a large number of sequential operations, can be very difficult. Thus, simulation for X propagation tracking is unreliable, as the simulation stimulus may not drive the RTL design into a state sequence that produces the X state condition. Similarly, the use of Linting tools and other mechanisms to find harmful X states often does not produce the desired result due to a variety of restrictions. In contrast, formal verification is ideal for X propagation analysis, given the exhaustive nature of the technique.

X propagation issues are further compounded by the nature of the HDL standards that can lead to "X Optimism." Standard Verilog, SystemVerilog, and VHDL will sometimes mask the propagation of X states through gated logic and other code where the path through the logic on which an X state would travel is disabled, a potentially optimistic scenario. For example, an AND function with a 0 on one input and an X on the other may have a simulated output of 0 (optimistic) or an X (pessimistic) depending on the simulation treatment of the design and the purpose of the simulation, see Fig. 5.7.

For some conditions it is desirable for the X state to propagate through a logic block if it appears on the input, regardless of the other inputs and the logic design. The X Optimism level of the verification analysis requires a level of controllability to ensure that a particular X state in question is evaluated correctly, given the design.

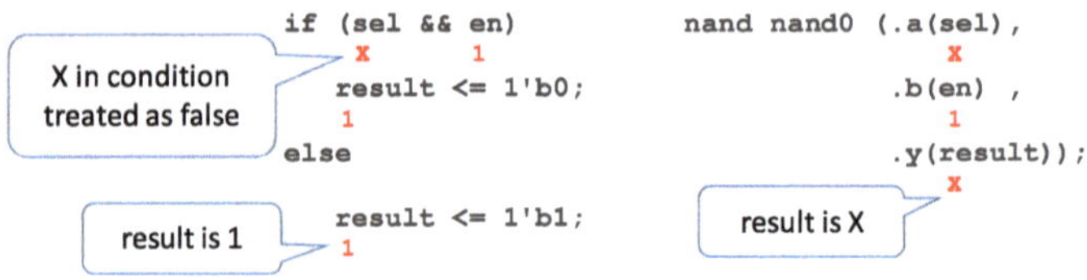

Fig. 5.7 X Optimism and pessimism examples

Many formal tools can be directed to create a model that has no X optimism and minimizes X pessimism, making it possible to analyze X propagation issues on the RT level.

Packaged formal apps can provide assertions to trap potential X state conditions, and many possible bugs may be found with relative ease. X Propagation Apps provide a robust and effective circuit analysis that highlights all the issues in a design that could lead to X state propagations. Because it is formal, it does not rely on simulation test stimulus generating the required coverage to find all X states. Some X prop apps use 4-state-logic formal analysis, allowing for easy debug and root cause analysis.

A typical X propagation app creates assertion sets that then are utilized by a formal engine to test for the reliable execution of reset and clock signals, examine design coding for common issues that lead to X states, check control logic that can mask common problems in other verification forms, plus other analysis options. Coding situations that could lead to an X state, include: divide by zero, array access attempt that could go out of the defined array bounds, uninitialized signals, case statements with incomplete assignments, and function that does return a value.

5.2.2.5 Data Transport Verification

In modern electronic systems, ensuring the correct transfer of data between and through components has become a complex challenge. The intermixing of multiple bus types that employ complex protocols, including Network-on-Chip (NOC) and bus bridge solutions, can often lead to a range of data errors. Furthermore, the preservation of data integrity in many design applications is of paramount importance. The verification of data transport using traditional simulation-based environments can often lead to transport issues not being identified, due to the likelihood that simulation stimulus will not provide an exhaustive test. As such, this is an ideal application for a formal-based solution, in this case a "Scoreboarding" App, where data transfer may be exhaustively checked without the need for simulation stimulus.

The use of Scoreboarding has become commonplace in many verification environments. Scoreboarding is an analysis function that tests for the correct transfer of data through electronic components of different types (Fig. 5.8). The mechanism has applications in many types of design with data transport activity, for example, communications designs making use of FIFO queues with logic to compare header and control word detail, bus systems of all types, including Network-on-Chip solu-

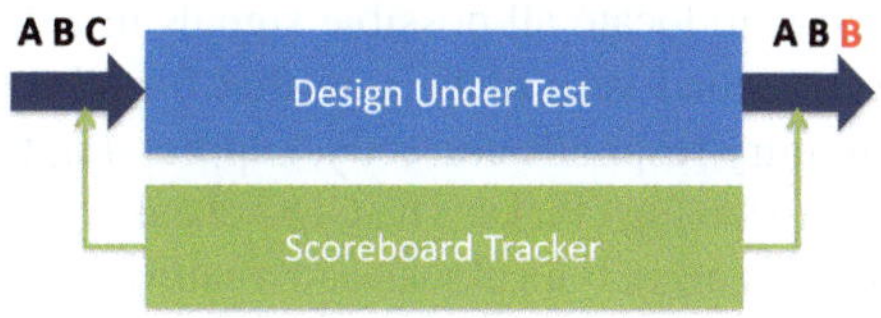

Fig. 5.8 Scoreboard tracking IO values

tions, data encoding and decoding blocks tested together to ensure no data blocks are corrupted, etc.

A Scoreboarding App can test for a number of data transport issues, including:

- Data corruption within a block
- Data misrouted or lost in some other fashion within a block
- Data duplication within a block to produce multiple outputs
- Data incorrectly reordered
- Data "ghosting" where data is apparently output without having been input
- Communication interruption or scrambling
- Ensuring that once data is input it will eventually appear on the output

The final issue on this list is particularly interesting as it highlights an advantage of formal over simulation. A Scoreboarding App can prove that a specific data word will eventually emerge from a block, a characteristic that may be recognized as a "liveness" property. Simulation-based solutions cannot make this guarantee unless the data emergence observation is actually made during a simulation run.

5.2.2.6 Clock Domain Crossing (CDC) Checks

As integrated circuits became larger, the clocking signals used to cycle the design logic became very difficult to manage. Delays in the clock lines would lead to signal timing issues and managing the clock tree around the device, lining the clocks up with the bus and other long connections became very unreliable. To maintain clock integrity, modern integrated circuits make use of multiple clock domains. Within the domain the clock signal is kept at a constant rate with minimal delays, and then each domain is treated as an asynchronous unit versus others in the device. There can be 100 s of domains in a modern SoC.

As these domains are asynchronous to each other, signals transitioning between the domains may be clocked by a sending domain, right before being again clocked by a receiving domain, or visa versa, creating signal glitches. This can lead to instability in the clocking flips flops causing circuit failure, and even oscillations. Circuitry has to be built in that handles glitches in the signals with respect to the domain clocks. Ensuring that all transitioning signals are handled correctly in all circumstances is a very complex task in simulation.

As such, one of the first formal apps to be produced was the so-called "Clock Domain Crossing" or CDC checker. This app, in general, has two functions. The first is to locate all possible signals in a design that could suffer from CDC issues. The second is to make sure that for all timing possibilities that the synchronizing circuitry responds correctly, suppressing the glitch for all combinations of sending and receiving clocks. The CDC checker essentially works by considering tiny timing increments in both clocks and treating them as separate states, analyzing the circuitry for its response to those states. Formal is ideal for this task as it can process these differing states in an exhaustive fashion.

The CDC checker has made safe clock domain usage possible in these large devices and is now extensively used by all semiconductor companies producing even reasonably small devices, right up to the largest of ICs.

5.2.3 Verification of Design Transformations

Throughout the design process, several transformations are applied to the design, including manual changes in order to optimize functions, as well as automated steps such as logic synthesis of RTL and place and route of netlists. There are different types of hardware bugs that may be introduced into an Integrated Circuit (IC). Design bugs introduced through human error during implementation are invariably eliminated during functional verification. Other systematic issues, on the other hand, like those introduced by the automated design refinement tools, are typically not directly checked by the functional verification process. These issues can be hard to detect, and damaging if they make it into the final device.

5.2.3.1 Logic Equivalence Checking

A primary example of formal verification has been logic equivalence checking, which has become a mandatory step in modern ASIC design flows. This application has become so popular, that often the terms formal verification and logic equivalence checking are used synonymously. Formal Equivalence Checking (EC) is used to eliminate synthesis errors in designs, by comparing the design with itself at different levels of abstraction. It is common to verify RTL input versus post synthesis gate level and P&R net list, as this must be "signed-off" before committing it to fabrication. This check, originally accomplished using simulation, eliminates systematic errors introduced by automated design refinement steps. The advent of formal verification-based Equivalence Checking for ASIC design provided an exhaustive comparison of the gate level to RTL code. As the RTL code has been verified, the overall solution provides a reliable method to commit the design to fabrication.

For ASIC synthesis, in general the synthesis tool converts register elements to flips flops and the combinational logic between these register elements is maintained as gates. EC tools that compare pre- and post- synthesis representations create a mapping between register elements and flips flops, and then compare the combination logic functionality in between.

Technically, logic equivalence checking verifies the correctness of design transformations of logic functions between a number of registers. In order to verify such transformations formally, first a mapping between inputs and outputs of two design representations, and a mapping between state holding elements of both representations is established. If the mapping is complete and the logic functions between these mapping points are proven to be equivalent, the whole of the two design representations is equivalent. A lot of research has been put into handling more and more logic design transformations efficiently, allowing for robust commercial solutions.

5.2.3.2 Sequential Logic Equivalence Checking

Logic equivalence checking cannot solve all design transformation verification tasks. In particular, it cannot handle sequential optimizations or other transformations that change the state representation of the design. For example, this method breaks down if the registers are changed between the representations, or moved relative to the combinational logic functional units.

The reason is that logic equivalence checking relies on the complete mapping of internal states. Once the state representation is changed, such a mapping can't be established and logic equivalence checking must fail. In this case, sequential logic equivalence checking is needed. A typical example for sequential design transformations and changes to the state representation is logic synthesis with re-timing and RTL design refactoring (Fig. 5.9).

Sequential equivalence checking mainly follows a conventional logic verification flow: design setup, mapping and comparison, and optional debug. However, the mapping of internal states is relaxed. In particular, the pairing of RTL to gate flops does not need to be complete to provide conclusive results. This is achieved by comparing the output behavior of the designs sequentially, starting from some initial state until a fixed point is reached, and no new design behavior can be observed. In case of nonequivalence the advantage is that, in contrast to logic equivalence checking, the sequential comparison allows the generation of counterexample traces

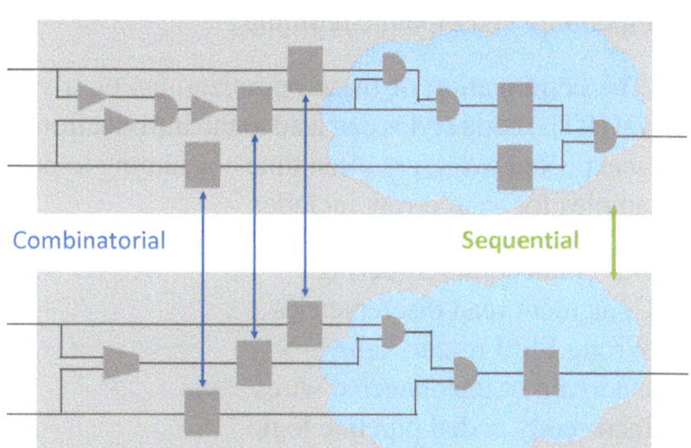

Fig. 5.9 Sequential versus combinatorial EC

for some compare points, independent of the mapping of other compare points. Consequently, mismatches can be debugged much more easily.

Once a block has been proven to operate correctly, a designer may wish to optimize some section, maybe to improve the coding style, reduce the gate count or streamline operation. As code is synthesized the resulting netlist might consume too much power or violate a timing constraint. Later in the process, an ECO may be required in a netlist without resynthesizing from the original RTL blocks. All these operations also benefit from a quick validation of functional equivalence. Without formal equivalence checking, an engineer must execute an entire simulation regression run for any change. This often requires a lot of time and may also need additional stimulus, with no guarantee that an exhaustive functional check will be performed. By formally comparing the overall functional relationship, sequential equivalence checking is able to exhaustively compare two descriptions for common functionality, pinpointing differences if they occur.

5.2.3.3 Sequential Equivalence Checking for FPGA Synthesis

FPGAs make use of a static hardware matrix, where the ratio of registers to inter-register logic is somewhat fixed. To drive the highest quality designs, state-of-the-art automated design flows leveraging aggressive optimizations are employed. Examples of such optimizations include:

- Register duplication and merging
- FSM re-encodings for power optimizations and functional safety
- TRM optimizations (triple modular redundancy)
- Generation of asynchronous feedback loops
- Use of IO cells and different bus resolution schemes
- Fixed gated clocks
- DSP block optimizations
- Shift register logic optimizations, including reset registers
- Use of FPGA RAM/ROM blocks, including Distributed- & Block-RAM
- Power optimizations, such as retiming
- Pipelining and pipeline retiming

The combination of these optimizations on a varied range of Register Transfer Level (RTL) code styles can lead to the introduction of errors. These errors can occur in unexpected ways, are often time consuming to detect, and potentially destructive. Examples for such errors include:

- Bus connection reordering
- Coincident read discrepancies
- Wrong FSM re-encoding
- Undriven or unconnected wires
- Incorrectly coded pipeline logic
- Incorrect IP parameter settings (Block RAMs, DSPs)

- Clock gating and low power issues
- Place and route connection issues
- Addition unspecified logic

Traditionally the only way to find these errors is heavy regression testing on the FPGA itself, with thorough testing that examines every possible operational scenario. This requires many days of stimulus creation and execution, with no guarantee of RTL to gate equivalence. Equivalence Checking is commonly used for ASIC design, but until recently it has not been possible in FPGA due to the nature of the sequential optimizations employed. It is now being leveraged for the same purpose in large Field-Programmable Gate Array (FPGA) designs.

5.3 Assertion-Based Verification of IP Blocks

Engineers often write directed tests that provide stimuli to a design-under-verification (DUV) to test different pieces of functionality and check that DUV responds as expected. Each test is directed at verifying a particular feature. As designs become more complex it becomes harder to cover all the possible scenarios and corner cases of the DUV. This is where typically advanced testbench automation comes into play to increase coverage and confidence. However, developing good constrained random testbenches and tuning them towards achieving very high coverage of the design functionality is anything but easy or a low-effort task. Therefore, many engineering teams have explored further enhancements in their verification flows, such as adding assertions and leveraging formal techniques. These have proven particularly effective for the verification of control and data transport blocks, as well as verifying complex scenarios hard to recreate during simulation.

5.3.1 Assertions in the Verification Flow

Over the past couple of years, assertion language standards, such as SystemVerilog Assertions (SVA) and the Property Specification Language (PSL) have been integrated into the major HDLs: SVA included in the Verilog standard in 2005, and PSL integrated into the VHDL 2008 standard. This standardization has been accompanied by a wide support of SVA or PSL in EDA tools, allowing engineers to reap significant benefits by adopting the new assertion features of the HDLs. While writing assertions requires additional effort, many engineering teams have recognized that they considerably shorten debug cycles by improving error localization and observability. Thus, adding a few simple assertions like "no FIFO overflow" or "no simultaneous read and write request" during the design phase of a block pays off later in the flow. Note that for adding such simple assertions, engineers do not need a deep knowledge of the intricacies of SVA or PSL. They can ramp up quickly because the coding of such simple assertions pretty much corresponds to RTL coding and documentation.

While it is widely accepted that assertions actually improve the verification flow, the major question is how to put assertions to best use. There are several aspects to this question, namely how to use assertions to verify nontrivial temporal aspects of the design and secondly, the question of whether adding a formal tool to achieve exhaustive verification of the assertions actually pays off. However, most leading semiconductor companies have deployed formal verification teams, clearly anticipating a significant return on investment.

5.3.1.1 Signal-Level Assertions

The major part of standards like SVA and PSL are tailored toward expressing temporal behavior; after all, simple combinational statements can be expressed in RTL. The main idea to express temporal behavior in SVA or PSL is borrowed from regular expressions, a concept well understood in computer sciences. However, assertions that express temporal behavior can be relatively cryptic and hard to understand. For example, what does:

“`req |=> (ack[=5] within (pending[*wait])) ##1 done`”

exactly mean? So while SVA provides the means to express complex temporal behavior, actually understanding how to code such SVA, and ensuring the assertion actual describes the intent, requires advanced training and significant experience.

An additional complication of temporal SVAs lies in debugging. The failure of a simple combinational assertion is easy to understand and to debug with common RTL debugging techniques and corresponding engineering skills. However, analyzing the failure of temporal assertions is an entirely different matter. Tools typically show the cycle where an assertion triggers, the cycles during which the assertion is “active”, and finally the cycle where it “fails”. Understanding why the displayed waveform violates the assertion is often a mind-twisting problem requiring a lot of expertise. After all, when writing temporal assertions, it is quite common that the first attempt of encoding the desired behavior was slightly incorrect. Hence, the engineer is presented with a counterexample that shows perfectly legal design behavior, but for some reason violates the assertion—maybe because the operator precedence does not correspond to the engineer’s expectations, changing the meaning of the assertion in an unintended, subtle way.

So in summary, while SVA and PSL were designed to express complex temporal behavior, significant training and hands-on experience is required to verify and debug temporal design aspects by assertions with cryptic regular expressions. In other words, using full-fledged PSL and SVA is not the best way to use assertions—it is simply too error prone and tedious.

5.3.1.2 RTL Style Assertions

In order to put assertions to better use, many engineering teams have combined the ease of use of simple combinational assertions with the familiar coding and debugging of RTL. Instead of using regular expressions to express temporal aspects, counters and FSMs are added to the verification code in plain RTL. For example, in order to distinguish legal input sequences from illegal ones, an FSM is encoded that "accepts" the legal input sequences and flags illegal sequences by entering an "error" state. After coding this in RTL, the corresponding assertion is once again a simple combinational one: just make sure that this FSM never enters the error state. So the combination of a simple assertion with some RTL in the verification code allows temporal behavior to be intuitively expressed. This really looks like the perfect solution as in addition to RTL coding knowledge, only minor SVA knowledge is required. Simulating such a temporal assertion has only a marginal performance overhead for the small piece of RTL code added to the simulation and debugging is similar to regular RTL code debug. Therefore, SVA with RTL for temporal aspects is much easier to adopt than full-fledged SVA/PSL.

There is one major drawback of the RTL-based solution, namely that the temporal behavior is really captured on a low, cycle-by-cycle execution level. In other words, while you may be able to "see" from full-fledged SVA that the sequence of events it covers is a "request", followed by 4 beats "valid", and finally an "acknowledge", the corresponding RTL version is much longer and features an FSM and some counters. It is easy to make mistakes using this RTL style, such as incorrectly encoding an FSM transition or forgetting to reset a counter.

So the use of RTL code within an assertion can make the assertion easier to understand and write, but is often inefficient and can lead to more mistakes, which are harder to detect.

5.3.1.3 Bug Hunting in Corner-Case Scenarios

A common use of formal techniques in today's designs is to test for a complex scenario and/or a corner-case operation that is hard to set up in simulation. As the design becomes large, attempting to drive the design state to a certain scenario using simulation stimulus can be a very long and laborious process. Essentially, the design must be reset and then executed, potentially over many cycles, to allow multiple sections of the design to get to a common point where some system-wide corner-case behavior is exhibited.

As previously noted, with formal an assertion maybe written that describes this corner-case directly, and asks the question "will the design fail in this situation?" As the formal tool has a representation of every possible state the design might transition through, the assertion will have the effect of testing exactly the right corner-case state, and exhaustively checking to ensure that the operation described will occur.

This use mode has been traditionally been described as "bug-hunting," where the engineer is looking for specific bugs around these corner-case states. However, this use model goes well beyond this description, ensuring that a broad range of operational scenarios occur correctly and verifying a very common source of bugs.

5.3.2 Verification Planning

For systematic assertion application, it is natural to encode functional requirements from the specification as assertions. For example, a protocol specification often lists a number of rules about the protocol inputs and outputs that are canonical candidates for temporal assertions. In addition, it is common to use assertions in a similar fashion to directed tests by anticipating operational corner cases and writing assertions that describe certain events that should never happen in these corner cases. A typical example is a counter that is not supposed to wrap from its maximum value back to zero. In this case an assertion may be included that tests the case where the counter is 0, no attempt is made to decrement it. This is similar to a directed test of the scenario "counter is empty", which has a fair chance of finding an issue with the wrapping counter if such an issue exists. In contrast to the functional requirements, these white-box assertions are not derived from the spec, but from knowledge of the RTL implementation and its corner cases. It is common that a large number of such assertions are developed, and it is also clear that some corner cases will not been covered by these assertions.

Beyond these typical assertions, methodologies widely vary in practice. Somehow, the general recipe is "the more assertions, the better". However, there is typically no way to assess what a newly added assertion contributes to the verification process, or if any additional scenarios are verified at all. Ideally, an engineer would like to systematically increase coverage by adding new assertions, and ideally know upfront how many assertions are needed in order to verify a certain block.

To sum up, it is quite common to develop a relatively large number of assertions for a block. These assertions do not systematically increase coverage, and it requires considerable expertise and insight into the RTL implementation in order to decide which assertions to develop.

In recent years, capacity and usability of formal tools has increased to such an extent that most hurdles for a wider adoption have been removed. However, the question still remains exactly how the results from a formal tool contribute to overall verification coverage. With simulation-based verification as the driving engine, it is hard to assess the impact of the exhaustive verification of some assertions to the overall verification status.

The EDA industry has spent significant effort in bridging the gap between formal tools and simulation coverage, for example by standardizing a universal coverage database API (Accellera's Unified Coverage Interoperability Standard or UCIS) designed to collect and collate results from dynamic and static tools. Even with this standard the actual impact of an exhaustively verified assertion on simulation coverage still cannot be quantified in meaningful and intuitive terms. This leads to a perception that adding formal analysis to the design and verification flow increases quality at the cost of increasing verification effort. It would actually be more desirable to save simulation effort by adding formal analysis, thus improving quality leveraging the same overall verification effort, or simply achieving the same verification quality faster.

The state of the art in verification planning is evolving considerably, particularly driven by the recent requirements of Safety Critical designs that must adhere to strict regulations. Planning techniques that break up an overall set of requirements, and then ensure that these requirements are individually tested by the most appropriate verification platform (simulation, formal, emulation, etc.) are commonplace. The back annotation of functional coverage information on to these individual tests is a key component of this process. The use of assertions often makes this process easier, and the coverage of these assertions is discussed in the next section.

5.3.3 *Quantitative Analysis and Coverage*

Verification Coverage has become a necessity, and is required to measure verification progress, and to what level the design has been covered. It is an increasingly critical component of modern verification environments. Unfortunately, the topic of effective models to measure accurate coverage remains hotly debated across the industry. Current coverage techniques providing approximations and insight as to verification progress and quality, but not an exact measure of "completeness." Most companies have settled on these metrics to monitor progress.

An ideal coverage measure would relate to functional coverage, or the level of which the design functionality has been covered. Testing the design verification against the specification would appear to provide such a metric, but who is to say the specification is complete? Nevertheless, functional coverage techniques are commonly used and these fit well with formal as assertions may be written to mimic specification detail.

Code coverage, that is the measurement of the verification of code elements, is in general a much easier metric to calculate, especially for simulation. While it is not a clear match to design functionality, it is generally accepted that if a high number of code elements have been tested, the verification process is of a high quality. Although code coverage is a common coverage mechanism used with simulation, it can be inherently harder to measure during formal verification than simulation, as the formal algorithm, in general, does not directly track code execution.

When considering formal coverage, it is important to differentiate between "controllability" and "observability". Controllability measures the effectiveness of the stimulus in activating aspects of the design functionality. Controllability is important in simulation, but is less of an issue in formal as the entire state space of the design is already essentially activated. However, it is possible to "over-constrain" the

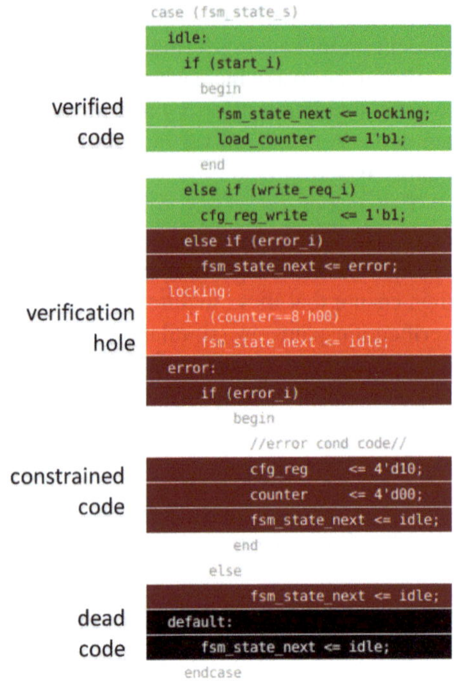

Fig. 5.10 Coverage classifications

verification run, which may mean that important design functionality does not get considered In the formal process, so this is still a factor for controllability coverage.

Observability is important for both simulation and formal, and measures how well possible issues are detected by the testbench checks, or assertions. Effective formal coverage is focused around observability metrics, ensuring that all functionality is covered by appropriate assertions.

Figure 5.10 shows different coverage classifications, including:

- Verified code: code that has been verified using at least one assertion
- Verification hole: code that is controllable and should be verified, but has not been covered by any assertions.
- Constrained code: code that is not controllable, i.e., a constraint has been provided that stops these lines from being considered as part of the verification process.
- Dead code: Code that can never be activated from an input. This is often caused by a design error, or in some cases is intentional

The primary formal coverage mechanisms in use today are as follows:

"Assertion Coverage" is a simple indication of assertion pass/failure, bounded proofs, and other quality metrics that may be passed back to a verification planning tool. Its limitation is that it only provides information on the status of the assertions, and not the coverage of the design code.

"Cone of Influence (COI) Coverage" is based on the activation of design logic that drives a particular signal. This is a somewhat conservative metric, which is reasonably easy to compute. However, it suffers from the accuracy issue that some of the design logic may be optimized differently in various formal engines, or may not be related to the design functionality of the covered design section.

"Proof Core" coverage is a variant on COI where the cone of influence is pruned down to focus the coverage effort on a specific section that actually drives the line of interest. This increases accuracy, but still suffers from some of the original COI issues, and is harder to calculate.

"Mutation Analysis" coverage is based on the concept is that if a testbench covers a specific code item, then a change in the code item should be detected by the testbench checks. As such, if each line of code in the design is slightly changed (or mutated) and the verification rerun, the difference in testbench behavior accurately indicates line coverage. It can be applied equally well to simulation and formal, producing a consistent coverage model between the two. It is very accurate but requires a lot of computation. Mutation analysis is used extensively in the software test space, and has become popular in hardware as well.

"Fault Observation Coverage" is a variant on Mutation Analysis. It uses the same principle and also provides a similar code coverage metric, which may be collated with simulation models. Instead of mutating the design itself, the design is unchanged, but a stuck at 1 or 0 fault is overlaid to hold signals to specific values, thereby modifying their behavior. If an assertion picks up this modified behavior, the signal or line is considered covered. Fault observation coverage is a model-based approach that directly works on the formal model rather than the original source code. This reduces the computation overhead of the algorithm by restricting its application to formal only.

5.4 Challenges Ahead

Semiconductor design is constantly changing, driven both by the continuous reduction in silicon geometries, as depicted by the famous Moore's Law, as well as new design requirements for applications made possible by the increase in computing power of these devices. Current applications for which new requirements are emerging including automotive electronics, Internet-of-things (IoT) technologies, advanced data processing compute farms, mobile applications, and others.

The use of Formal Verification has an impact on a number of the necessary requirements for these market segments, and in some cases is seen as a necessity. Key design flow trends for which formal technology is ideal includes:

(a) High-Level Design
(b) High Reliability & Safety
(c) HW Security
(d) Low Power Devices

5.4.1 High-Level Design

High-Level Synthesis (HLS) (aka behavioral synthesis) transforms algorithmic and potentially untimed design models often written in SystemC and C++ to fully timed Register Transfer Level (RTL) design blocks. HLS tools are particularly popular as a method to rapidly generate design components with varying microarchitectures, while optimizing algorithm-processing datapaths rapidly and effectively. This provides substantial benefits in terms of flexibility and time-to-market. Consequently, HLS is now in use at many large semiconductor and electronic systems companies, particularly those involved with applications that are heavily Digital Signal Processing (DSP), Image Processing, or Communications Processing oriented.

However, the verification options for SystemC and C++ designs have not kept pace with the synthesis technology. Simulation-style verification of HLS code is largely performed by compiling and debugging the design representation, linked with the Accellera OSCI SystemC Class Library, in a similar fashion to software test. Due to the limited availability of SystemC tools, much of the verification task is performed on the resulting synthesized RTL code, introducing a level of indirection that makes correcting issues at the SystemC/C++ level complex and time consuming.

The primary verification requirement is to allow thorough verification of algorithmic code prior to synthesis, in order to ensure that the abstract algorithm implementation is tested and fully optimized against the original specification, as well as avoiding long debug cycles. In addition, artifacts of the SystemC standard, for example the lack of an unknown, or X, state, potential race conditions between threads, etc., result in further ambiguity that must be eliminated before synthesis. Specific issues related to this abstract design level may be easily tackled with the right verification methods, improving final design quality.

While formal verification is not an ideal approach for design blocks that are performing heavy arithmetic computation, it is extremely useful in solving the above issues. Hardware issues, such as the detection of uninitialized value propagation, or the effects of undefined operations (e.g., array out-of-bounds) may be effectively detected using formal automated apps at the SystemC level. In addition, algorithmic-specific verification tasks can also be addressed using formal. A good example is the automated arithmetic overflow check, which allows for the inspection of number precision across algorithmic datapaths.

5.4.2 High Reliability and Safety Critical Systems

The use of "fail-safe" electronic components in safety critical systems is now commonplace through many electronics industry sectors, including Automotive, Aerospace, Power Generation, Defense, and Medical devices. Governed by a range of regulatory standards, the verification of these systems, in general, must be proven to be as rigorous as possible. In addition, to guarantee the safe operation of these

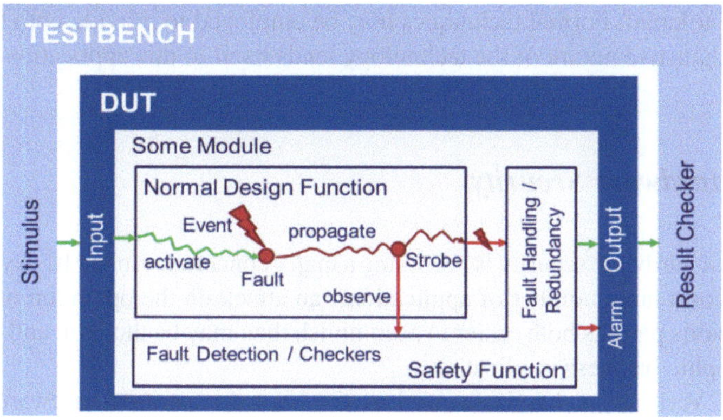

Fig. 5.11 Fault propagation in a safety critical design

systems, safety mechanisms are integrated that ensure a reliable, deterministic reaction to random hardware failures when the device is operating in the field. These too must be verified to operate correctly and trap operational hardware faults.

It is hard to demonstrate that simulation-only verification solutions can provide the required degree of coverage necessary to guarantee safety. The exhaustive nature of formal verification solutions makes it a natural fit for these designs. However, additional capability must be included to prove design reliability and failsafe operation. Design reliability can be shown by utilizing advanced coverage techniques to demonstrate that any "systematic" bug (a bug introduced during the development process) in the design would indeed be detected by a series of assertions, executed on a formal platform.

To validate safety mechanisms that trap and resolve "random" field problems, the ISO 26262 and other standards demand a quantitative analysis of random hardware failures and their outcomes. Part of this analysis comprises the injection of faults into the gate level models of integrated circuits during verification to prove that faults will be detected by a safety function (Fig. 5.11). These gate-level models can be complex and contain numerous possible fault scenarios.

Formal Verification clearly has application in this area. The discovery of Systematic issues essentially requires the use of effective verification practices described throughout this chapter, together with a high degree of coverage measurement to ensure quality.

Random errors, that is those that occur during the operation of the device due usually to external affects, require the implementation of special capability on the design itself to trap and eliminate the effect of these errors. It is important that the operation of these circuit elements is also verified, and this requires the injection of faults around the design, and observing that the faults cannot propagate beyond the safety mechanisms. In addition to this verification process, diagnostic coverage must also be performed to record the proportion of design structures that have been tested to be fault tolerant. Formal techniques may be employed to provide both functions, and the exhaustive nature of the technology lends itself to this application.

5.4.3 *Hardware Security*

The issue of hardware security is becoming a major concern for many IC development projects. For a large number of applications, an attack on the operation of a device by a malicious party is both easier to accomplish than may be thought, and can result in catastrophic business implications.

Formal Verification has been used as the basis for various hardware security testing tools that can test for several different security properties, such as isolation, noninterference, and the presence of timing/digital side channels. These tools can ensure that both security confidentiality and integrity are being enforced on any given part of your hardware design, as follows:

Confidentiality—The secrecy of a cryptographic key is often the crux of the security scheme. Ensuring that this key is kept confidential from less trusted parts of the chip is an important area of concern. Formal security tools can prove it can never flow to any part of a system that is designated as "untrusted" (Fig. 5.12). This means that, regardless of what software is running on the system, the key is provably safe from being leaked.

Integrity—For many applications, a common device may be responsible for the operation of both critical and noncritical components. For instance, in new automotive vehicles, the same system may be responsible for the operation of the brake system as well as the satellite radio. Formal security tools can test for different noninterference

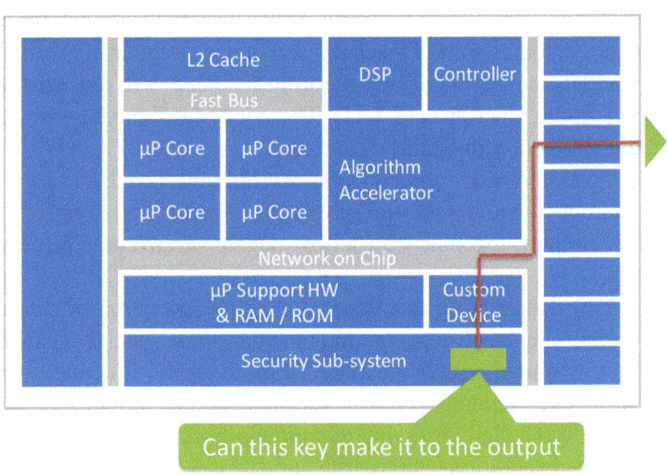

Fig. 5.12 Checking for protected key vulnerabilities

properties to prove that one part of the system will neve
part, whether it is through explicit changes in values or di

5.4.4 Low-Power Devices

The subject of low-power has become ubiquitous in electr
requiring some level of power control. The obvious and m
mobile and other battery operated devices, or systems that
of reduced energy such as solar power. However, all elec
from sensible power control as it can only improve the
particular electronic sector where low powered devices a
that of the "Internet of Things" or IoT. It has been assum
will require remote sensors that are able to record some i
wirelessly to a central station, all on the power of a solar
year battery. These applications may well drive a new leve
electronic systems.

There are many strategies for reducing the power con
these can be controlled both from the system software o
of the most common is the powering down of sections of
for example ARM's "Big Little" processor configuration
processor for general device operation, but can switch on a
of heavy data processing.

The switching in and out of sections, or "power domai
complex business. The power domain must be activated and b
levels, reset, and then connected into the system while the rest
to function without missing a step. This process may vary cons
the status of other parts of the system and the operation to whic
domain will be deployed, and this in turn leads to a lot of oper
to be verified.

Formal techniques are being used to test the power up and corr
of power domains. Apps previously described such as X propagati
in this regard, to ensure that unwanted artifacts of a power doma
do not corrupt the smooth operation of the rest of the device. C
be performed on control communication sequences and register ope
power domain is running, to ensure that the right information is pr
domain status is correct in any circumstances.

It is unclear if IoT will bring new requirements to this field. Certain
problem is getting worse and more sophisticated control mechanisms n
degrees of verification. One change that has occurred is more control fron
layers of the software protocol stack through a control processor, which h
previous hardware control, and somewhat reduced the effectiveness of ea
standards such as the Unified Power Format (UPF).

ıcation tool and service vendor websites. OneSpin Solutions: www.onespin.com Ca- n Systems: https://www.cadence.com/content/cadence-www/global/en_US/home/ n-design-and-verification/formal-and-static-verification.html Synopsys: https:// ›sys.com/verification/static-and-formal-verification.html Mentor Graphics: https:// r.com/products/fv/questa-formal/ IBM: https://www.research.ibm.com/haifa/ ification/Formal_Methods-Home/ Oski Technology: http://www.oskitechnology.

:ld.org: A website dedicated to formal verification, http://www.formalworld.org

, Test and verification formal day conference, http://www.testandverification.com/ ɛs/formal-verification-conference/, 2013-'16

Electronic Systems Design, University of Berkley. Introduction to formal verification, ıbedded.eecs.berkeley.edu/research/vis/doc/VisUser/vis_user/node4.html

, How engineers feel about formal verification: Dac report #1, http://www.deepchip. ns/dac16-01a.html, 2016

ı, Formal verification primer, http://www.deepchip.com/items/0558-01.html, 2016

hal, These five principles define formal verification, http://electronicdesign.com/eda/ ve-principles-define-formal-verification, 2015

Biographies

inkmann co-founder of OneSpin Solutions, was named president and CEO in May 2012. gs to this role more than 15 years of broad expertise in management, verification and al R&D, having served at companies in both Europe and the U.S. Previously, as vice nt of OneSpin's software development, he successfully turned the company's fundamen- ntific and engineering innovations into state-of-the-art, award-winning EDA tools. Prior to ›in, Brinkmann worked in the formal verification group at Infineon Technologies in Mu- Previously, at Siemens, he conducted industrial research projects on formal verification at the any's Corporate Research Labs in the U.S. and Germany. He began his career in the systems cation group at Siemens Public Communication Networks, and later focused on hardware ıation, formal verification and systems-on-a-chip design. Dr. Raik Brinkmann holds a Diplom rmatiker from the Clausthal Technical University, Germany and a Dr.-Ing. (equivalent to a). degree) from the Department of Electrical Engineering at the University of Kaiserslautern, rmany.

ıve Kelf is the Vice President of Marketing at OneSpin. Previously to OneSpin he was president and CEO of Sigmatix, Inc. He worked in various roles in sales and marketing at Cadence esign Systems, and was responsible for the Verilog and VHDL verification product line. As vice resident of marketing at Co-Design Automation and then Synopsys, Kelf oversaw the successul introduction, standardization and growth of the SystemVerilog language, before running marketing for Novas Software, the hardware debug solution leaders, which became Springsoft (now Synopsys). With his strong engineering background, he has also held positions in various industry bodies, including the Treasurer of Accellera. Kelf holds a Master of Science degree in Microelectronics from Brunel University of West London and an MBA from Boston University.

MIX
Papier aus verantwortungsvollen Quellen
Paper from responsible sources
FSC® C105338

If you have any concerns about our products,
you can contact us on
ProductSafety@springernature.com

In case Publisher is established outside the EU,
the EU authorized representative is:
Springer Nature Customer Service Center GmbH
Europaplatz 3, 69115 Heidelberg, Germany

Printed by Libri Plureos GmbH
in Hamburg, Germany